Latest Advances in Preservation Technology for Fresh Fruit and Vegetables

Latest Advances in Preservation Technology for Fresh Fruit and Vegetables

Editor

Peng Jin

MDPI • Basel • Beijing • Wuhan • Barcelona • Belgrade • Manchester • Tokyo • Cluj • Tianjin

Editor
Peng Jin
Nanjing Agricultural
University
China

Editorial Office
MDPI
St. Alban-Anlage 66
4052 Basel, Switzerland

This is a reprint of articles from the Special Issue published online in the open access journal *Foods* (ISSN 2304-8158) (available at: https://www.mdpi.com/journal/foods/special_issues/preservation_technology_fruit).

For citation purposes, cite each article independently as indicated on the article page online and as indicated below:

LastName, A.A.; LastName, B.B.; LastName, C.C. Article Title. *Journal Name* **Year**, *Volume Number*, Page Range.

ISBN 978-3-0365-6475-3 (Hbk)
ISBN 978-3-0365-6476-0 (PDF)

Contents

About the Editor

Peng Jin

Jin Peng is a PhD Professor, and PhD Supervisor. In June 2009, he graduated from Nanjing Agricultural University with a Ph.D. degree in Food Science, and worked as a Teacher in the College of Food Science and Technology, mainly engaged in the research of post-harvest biological process, storage and preservation technology of fruits and vegetable. From September 2007 to September 2008, he visited the Agricultural Products Quality and Safety Laboratory (PQSL) of the United States Department of Agriculture to conduct research on the post-harvest preservation and storage technology of fruits and vegetables. In 2011, he was selected as the "Zhongshan Academic Rookie" of Nanjing Agricultural University. In 2012, he was selected as the training target of outstanding young backbone teachers of the "Blue Project" in the Jiangsu Province. His research focuses cover the research on post-harvest chilling damage, disease control and fresh-keeping technology of southern characteristic fruits (loquat, bayberry, peach, strawberry, grape, etc.). He has chaired three projects of the National Natural Science Foundation of China, participated in one project of the National "Twelfth Five-Year" Science and Technology Support Plan, and one project of the National Public Welfare (Agricultural) Industry Special Project. He has published more than 60 papers on post-harvest diseases and preservation techniques of fruits and vegetables in Chinese and international journals, of which 45 are indexed by SCI (15 as the first author) and 5 by EI.

foods

MDPI

Editorial

Latest Advances in Preservation Technology for Fresh Fruit and Vegetables

Peng Jin

College of Food Science and Technology, Nanjing Agricultural University, Nanjing 210095, China; pjin@njau.edu.cn

Fruit and vegetables contain abundant nutrients, as well as dietary and health benefits, and economic value, but suffer from shorter shelf life, declining quality, and rapid deterioration after harvest. Several preservation technologies can be used with fresh produce, such as chemical treatments, physical applications, and other approaches. This Special Issue aimed to collect the latest trends in diverse mechanisms of fresh product preservation and covered six chemical treatments: 1-Methylcyclopropene (1-MCP) [1], calcium chloride ($CaCl_2$) [2], melatonin [3], hydrogen sulfide (H_2S) [4], essential oils [5], lactic acid and acetic acid [6], in a broader postharvest quality attribute. Other treatments, such as cold shock, modified atmospheric packaging [7], and chlorophyllin-based photodynamic inactivation [8], were utilized in postharvest conditions.

1-MCP plays a crucial role in postharvest products, suppressing respiration and limiting ethylene peak. The study reported by Dai et al. showed that 100 $\mu L\ L^{-1}$ concentration of 1-MCP treatment effectively induced antioxidant capacity in Chives (*Allium schoenosprasum* L.). The application of 1-MCP in this medicinal leafy vegetable considerably improved organosulfur profiles such as isoalliin and total-S-alk(en)ylcysteine sulfoxides (ACSOs) content [1]. The positive correlation between ACSOs and antioxidant enzyme activity is crucial in preserving the quality. Thus, it demonstrates the potential function of 1-MCP in maintaining quality while being stored [1]. In other conventional chemical treatments, Hou et al. studied the possible mechanism of $CaCl_2$ in loquat (*Eriobotrya japonica*) fruit postharvest storage maintenance [2]. Several shreds of evidence show that Ca^{2+} is an essential trigger in the stress tolerance physiology of plants. The study showed that $CaCl_2$ mitigated chilling injury and improved quality by modulating reactive oxygen species (ROS) homeostasis via higher antioxidant activities and scavenging capacity. Similarly, $CaCl_2$ improves the ascorbate-glutathione (AsA-GSH) cycle and a higher expression level of corresponding genes, namely ascorbate peroxidase (*EjAPX*), glutathione reductase (*EjGR*), monodehydroascorbate reductase (*EjMDHR*), and dehydroascorbate reductase (*EjDHAR*) [2]. This explains the cold tolerance induced by $CaCl_2$ treatment in loquat fruit.

As an organic preservative, melatonin is a vital stress regulator in plants physiology, with a possible role in the postharvest life of fruit and vegetables. The application of melatonin (600 $\mu mol\ L^{-1}$) remarkably improved quality in postharvest guava (*Psidium guajava* L.) fruit [3]. The mechanisms involve enhancing enzymatic and non-enzymatic antioxidants, paralleled with regulating lipid metabolisms and defense-related enzymes such as chitinase, phenylalanine ammonia lyse, and 4-coumaric acid-CoA-ligase. It further helps maintain cellular integrity and boosts disease resistance, which results in the improvement of storage quality in guava [3]. In another gaseous preservative, H_2S is utilized in horticultural products with low concentrations. The study proves that H_2S effectively improved antioxidant activity and increased the activities of enzymes involved in energy metabolisms, including cytochrome C oxidase (CCO), succinate dehydrogenase (SDH), H^+-ATPase, and Ca^{2+}-ATPase [4]. Furthermore, H_2S regulated proline metabolisms by enhancing key synthesizing enzymes, ornithine aminotransferase (OAT) and Δ^1-pyrroline-5-carboxylate syntheses (P5CS) activities. Therefore, H_2S could potentially alleviate chilling

Citation: Jin, P. Latest Advances in Preservation Technology for Fresh Fruit and Vegetables. *Foods* **2022**, *11*, 3236. https://doi.org/10.3390/foods11203236

Received: 19 September 2022
Accepted: 14 October 2022
Published: 17 October 2022

injury and maintain the quality of cucumber fruits [4]. In another study, an essential oil P-Anisaldehyde (PAA) was used for postharvest fruit preservation due to its antimicrobial properties [5]. The treatment of PAA improved the firmness, total soluble solids, total phenols, and flavonoids in postharvest pitaya fruit (*Hylocereus undatus*). In addition, hydrogen peroxide and superoxide anion production were limited by the essential oil treatment. Thus, the maintenance of fruit quality was attributed to the activated antioxidant enzymes namely superoxide dismutase (SOD), peroxidase (POD), catalase (CAT), and higher expressions of several corresponding genes encoding antioxidants enzymes (*HpSOD*, *HpPOD*, *HpCAT*, *HpAPX*, *HpGR*, *HpDHAR*, and *HpMDHAR*) [5]. In the proteomic study, 0.8% lactic acid + 0.2% acetic acid treatment was used to inhibit inhibit *Escherichia coli* O157:H7 contamination in fresh products [6]. This treatment managed to reduce membrane damage via a positive alteration of the stress response, catabolism, and molecular activities. The additional investigation suggested that the combined treatment of lactic acid and acetic acid could be used to preserve *E.coli*-susceptible fruit and vegetables [6].

Food waste postharvest poses a significant bottleneck, and an innovative reused approach is crucial in these areas. The antimicrobial properties of the 12 discarded vegetables were studied for distractive soft rot disease [9]. In tested vegetables, tomatoes showed a higher growth of bacteriocin—*Lactobacillus paracasei* WX322, which was induced by 10 days of fermentation of tomato juice. This bacteriocin is an antibacterial (precursor) polypeptide. The *Lactobacillus paracasei* WX322 causes severe damage to soft rot pathogen, namely *Pectobacterium cartovorum* (*Pcb* BZA12). Eventually, antimicrobial activity reduced soft rot and decay in cucumber, tomato, and green beans [9]. Therefore, innovative value-added organic preservatives could be applied to postharvest fruit and vegetables. In another disease-response-related report, the peptide PAF 56 (GHRKKWFW) shows a control mechanism in economically important citrus pathogenic fungi (*Penicillium digitatum*, *Penicillium italicum* and *Geotrichum citri-aurantii*) [10]. The pathways of action of the peptide on the spores of pathogen are primarily ascribed to the regulation of time-dependent cell membrane permeability and degree of membrane damage [10]. This provides insights for future research applications and commercial use.

In physical treatment approaches, the study showed that the combined effect of cold shock and passive atmosphere packaging (CS-PAP) significantly improved the quality and sensory attributes, as well as the nutritional value in postharvest cucumber fruit (*Cucumis sativus* L.) [7]. Thus, a related physical intervention could be applied for potential preservation under postharvest conditions [7]. Concerning another physical treatment, chlorophyllin-based photodynamic inactivation (Chl-PDI) affects the storage quality of pakchoi (*Brassica rapa subsp. chinensis*) [8]. The report demonstrated that chlorophyllin (1×10^{-5} mol L^{-1} and 405 nm light (22.27 J cm^{-2} per day)) treatment improved quality indicators, including vitamin C, total soluble solids and color preservation. Furthermore, it eliminated ROS accumulation by enhancing SOD and POD enzymes activities. The chlorophyllin treatment possibly attacked bacterial extracellular polysaccharides and extracellular proteins in vegetables. Hence, Chl-PDI is effectively used for fresh-cut preservation and displays good antimicrobial properties [8].

This editorial summarizes the selected latest chemical preservatives applications on fruits and vegetables for the regulation of antioxidant attributes, defense response, cold stress regulation, and subsequent shelf-life extension. Similarly, physical treatment can be applied to storage for preservation of fruits and to reduce distractive disease infestation.

Funding: This research received no external funding.

Institutional Review Board Statement: Not applicable.

Conflicts of Interest: The author declares no conflict of interest.

References

1. Dai, X.; Lu, Y.; Yang, Y.; Yu, Z. 1-Methylcyclopropene Preserves the Quality of Chive (*Allium schoenoprasum* L.) by Enhancing Its Antioxidant Capacities and Organosulfur Profile during Storage. *Foods* **2021**, *10*, 1792. [CrossRef] [PubMed]
2. Hou, Y.; Li, Z.; Zheng, Y.; Jin, P. Effects of $CaCl_2$ Treatment Alleviates Chilling Injury of Loquat Fruit (*Eriobotrya japonica*) by Modulating ROS Homeostasis. *Foods* **2021**, *10*, 1662. [CrossRef] [PubMed]
3. Fan, S.; Xiong, T.; Lei, Q.; Tan, Q.; Cai, J.; Song, Z.; Yang, M.; Chen, W.; Li, X.; Zhu, X. Melatonin Treatment Improves Postharvest Preservation and Resistance of Guava Fruit (*Psidium guajava* L.). *Foods* **2022**, *11*, 262. [CrossRef] [PubMed]
4. Wang, J.; Zhao, Y.; Ma, Z.; Zheng, Y.; Jin, P. Hydrogen Sulfide Treatment Alleviates Chilling Injury in Cucumber Fruit by Regulating Antioxidant Capacity, Energy Metabolism and Proline Metabolism. *Foods* **2022**, *11*, 2749. [CrossRef] [PubMed]
5. Xu, Y.; Cai, Z.; Ba, L.; Qin, Y.; Su, X.; Luo, D.; Shan, W.; Kuang, J.; Lu, W.; Li, L.; et al. Maintenance of Postharvest Quality and Reactive Oxygen Species Homeostasis of Pitaya Fruit by Essential Oil *p*-Anisaldehyde treatment. *Foods* **2021**, *10*, 2434. [CrossRef] [PubMed]
6. Wang, J.; Lei, Y.; Yu, Y.; Yin, L.; Zhang, Y. Use of Acetic Acid to Partially Replace Lactic Acid for Decontamination against *Escherichia coli* O157:H7 in Fresh Produce and Mechanism of Action. *Foods* **2021**, *10*, 2406. [CrossRef] [PubMed]
7. Wang, F.; Mi, S.; Chitrakar, B.; Li, J.; Wang, X. Effect of Cold Shock Pretreatment Combined with Perforation-Mediated Passive Modified Atmosphere Packaging on Storage Quality of Cucumbers. *Foods* **2022**, *11*, 1267. [CrossRef] [PubMed]
8. Zhang, Y.; Ding, Z.; Shao, C.; Xie, J. Chlorophyllin-Based 405 nm Light Photodynamic Improved Fresh-Cut Pakchoi Quality at Postharvest and Inhibited the Formation of Biofilm. *Foods* **2022**, *11*, 2541. [CrossRef] [PubMed]
9. Zhu, R.; Liu, X.; Li, X.; Zeng, K.; Yi, L. Transformation of Inferior Tomato into Preservative: Fermentation by Multi-Bacteriocin Producing *Lactobacillus paracasei* WX322. *Foods* **2021**, *10*, 1278. [CrossRef] [PubMed]
10. Wang, W.; Feng, G.; Li, X.; Ruan, C.; Ming, J.; Zeng, K. Inhibition of Three Citrus Pathogenic Fungi by Peptide PAF56 Involves Cell Membrane Damage. *Foods* **2021**, *10*, 2031. [CrossRef] [PubMed]

foods

MDPI

Article

1-Methylcyclopropene Preserves the Quality of Chive (*Allium schoenoprasum* L.) by Enhancing Its Antioxidant Capacities and Organosulfur Profile during Storage

Xiaomei Dai [1,2], Yaping Lu [3], Yuan Yang [1] and Zhifang Yu [1,*]

1 College of Food Science and Technology, Nanjing Agricultural University, Nanjing 210095, China; 2019208001@njau.edu.cn (X.D.); 2020808125@stu.njau.edu.cn (Y.Y.)
2 Department of Food Science and Technology, Jiangsu Food & Pharmaceutical Science College, Huaian 223003, China
3 College of Life Science, Nanjing Agricultural University, Nanjing 210095, China; lyphwq@njau.edu.cn
* Correspondence: yuzhifang@njau.edu.cn; Tel.: +86-25-84399098

Abstract: The quality, antioxidant capacities, and organosulfur profile of chives (*Allium schoenoprasum* L.) treated with 1-methylcyclopropene (1-MCP) during storage were investigated in this study. The 1-MCP treatment (100 μL/L, fumigation 12 h at 20 °C) effectively inhibited tissue respiration and H_2O_2 production, enhanced the ascorbic acid (ASA) and glutathione (GSH) content, and promoted the activity of antioxidant enzymes (superoxide dismutase SOD, Catalase CAT, and ascorbic peroxidase APX) during the 5-day storage period at 20 °C. The result further showed that the 1-MCP treatment inhibited chlorophyll degradation, alleviated cell membrane damage, and delayed the chive senescence, with the yellowing rate being reduced by 67.8% and 34.5% in the 1-MCP treated chives on days 4 and 5 of storage at 20 °C, respectively. The free amino acid content of the chive was not affected by the 1-MCP treatment at 20 °C. However, the senescence rate of the chive was not reduced by the 1-MCP treatment when stored at 3 °C. The liquid chromatography data further showed that the 1-MCP treatment induced a 15.3% and 13.9% increase in the isoalliin and total S-alk(en)ylcysteine sulfoxides (ACSOs) content of the chive on day 2 at 20 °C, respectively. Furthermore, there was a strong positive correlation between ACSOs content and CAT/APX activity, indicating that ACSOs probably played a key role in enhancing the antioxidant capacities of the chive during storage at 20 °C. Thus the study efficiently demonstrates that 1-methylcyclopropene preserves the quality of chive (*Allium schoenoprasum* L.) by enhancing its antioxidant capacities and organosulfur profile during storage.

Keywords: chive; senescence; quality; antioxidant capacity; amino acids; organosulfur compounds

Citation: Dai, X.; Lu, Y.; Yang, Y.; Yu, Z. 1-Methylcyclopropene Preserves the Quality of Chive (*Allium schoenoprasum* L.) by Enhancing Its Antioxidant Capacities and Organosulfur Profile during Storage. *Foods* **2021**, *10*, 1792. https://doi.org/10.3390/foods10081792

Academic Editor: Michael Kontominas

Received: 11 June 2021
Accepted: 29 July 2021
Published: 2 August 2021

1. Introduction

The *Allium* genus includes several hundred species and is widely utilized as a drug in folk medicine for its potent antibacterial, hypoglycemic, hypolipidemic, hypocholesterolemic, cardiovascular, antithrombotic, and antitumor activities [1–4]. *Allium schoenoprasum* L (chive) is rich in carbohydrates, proteins, amino acids, vitamins, and minerals and is cultivated as vegetables or seasoning herbs all around the world [5]. Like other *Allium* plants, chives have a unique flavor, and the precursor for flavor and therapeutic compounds are S-alk(en)ylcysteine sulfoxides (ACSOs) [6].

Similar to many green leafy vegetables, postharvest chives have a high respiration and senescence rate which often leads to quality deterioration, including leaf yellowing, wilting, and nutrient loss at ambient temperature storage. The most obvious characteristics of senescence are degradation of chlorophyll, breakdown of chloroplasts, increased oxidation, and hydrolysis of macromolecules, such as proteins [7]. During postharvest storage, leaf senescence is induced by different stress conditions, which culminate in an increased level of reactive oxygen species, ROS (including O_2^-, H_2O_2, and $\cdot OH^-$) [1] and its concomitant oxidative reaction. In addition to their destructive effect on cells, ROS can act as signaling molecules, by enhancing the antioxidant system for scavenging ROS [8]. When ROS homeostasis is broken down, membrane-lipid or protein oxidative damage occurs [9]. Research on the senescence mechanism of chive is very important for prolonging the shelf life of the *Allium* genus plant.

The current consensus is that the postharvest treatment-dependent delay in leaf senescence in the vegetables is mostly attributed to decreased ROS production [10]. To date, postharvest 1-MCP treatment has proven to be effective in extending the postharvest life of a wide range of vegetables, such as Chinese kale [11], broccoli [12], and pak choi [13]. The mechanism of 1-MCP delay in senescence of vegetables is not only by inhibiting ethylene action but also by enhancing stress response and defense [14], such as enhancing antioxidant enzyme activity [15]. The effect of 1-MCP treatment on the postharvest quality of the *Allium* genus has been rarely reported in the literature. Thus, a study on the effect of 1-MCP treatment on the storage quality of chive will be practically useful in the postharvest management of the *Allium* genus plant.

The physiological and biochemical properties of similar plant species such as garlic and onion have been extensively studied [16–20]. However, there are few literature reports on the relationship between S-alk(en)ylcysteine sulfoxides (ACSOs) and antioxidant capacity in the *Allium* plant during storage. The objective of this study is to determine the impact of 1-MCP treatment on leaf quality, antioxidant capacity, and organosulfur compounds of chive during storage at temperatures (20 °C and 3 °C). The relationship between ACSOs and the antioxidant capacity of the chive leaf was also studied. The results of this work will provide insight into the mechanisms of senescence physiology in *Allium* vegetables as well as the role of ACSOs in *Allium* vegetable senescence.

2. Materials and Methods

2.1. Materials and Treatments

The chive (*Allium schoenoprasum* L. cv. Xinghua), grown in Zhujia village, Xinghua County (nationally the most condensed cultivating area), Jiangsu Province, China, was used for this study. Harvested chives (day 0) were transported to the laboratory within 3 h at 20 °C. The senescent leaves were removed while the leaves with uniform sizes were selected and dried with a fan. The sample was then separated into four groups, CK 20 °C (non-treated), CK 3 °C (non-treated), 1-MCP 20 °C and 1-MCP 3 °C (treated with 1-MCP).

In the preliminary study, the chive leaves were fumigated with 1, 5, 10, 50, 100, 200 and 500 μL/L 1-MCP for 12 h at 20 °C and stored for 5 d at 20 °C with 85–90% humidity. The results indicated that 100 μL/L 1-MCP treatment was the most effective in delaying senescence in the leaf. Thus, 100 μL/L 1-MCP was used to fumigate the sample for 12 h at 20 °C. Every 4 kg chives were fumigated in a plastic container. A Petri dish containing 1% (*w*/*v*) KOH aqueous solution was kept inside the plastic container to prevent the accumulation of CO_2 by chives. After fumigation (day 0.5), the CK 20 °C group and 1-MCP 20 °C group were stored at 20 °C (RT) with 85–90% humidity for 5 d, while the CK 3 °C group and 1-MCP 3 °C group were stored at 3 °C (LT) with 85–90% humidity for 20 d (Table 1).

Table 1. Treatments and storage conditions of chive vegetables in four groups.

Group	Treatment	Storage Condition	Sample Time
CK 20 °C	No	20 °C	Day 0, 0.5, 2, 4 and 5
1-MCP 20 °C	1-MCP	20 °C	Day 0, 0.5, 2, 4 and 5
CK 3 °C	No	3 °C	Day 0, 0.5, 2, 5, 8, 12, 16 and 20
1-MCP 3 °C	1-MCP	3 °C	Day 0, 0.5, 2, 5, 8, 12, 16 and 20

Note: Day 0 means the day on harvest; Day 0.5 means at the time when 1-MCP fumigation for 12 h was just completed.

At every testing point, 3 kg chives were sampled in each group in three biological replicates (1 kg chives per replicate) for quality parameter analysis, physiology and ROS level. After the above determinations, samples of size 0.5–1 cm were collected, immediately frozen in liquid nitrogen, and stored at −80 °C for further analysis. As the white shaft part was very short, chives were sampled starting from the white and green transition to 2 cm of the top of the leaves, which represent the whole plant.

2.2. Reference Compounds

S-methyl-L-cysteine sulfoxide (MCSO, CAS:6853-87-8, HPLC $\geq$ 98%) was purchased from Nanjing Chemlin Chemical Industrial Co., Ltd.; S-allyl-L-cysteine sulfoxide (ACSO, CAS:556-27-4, HPLC $\geq$ 98%) was purchased from Shanghai Yuanye Bio-Technology Co., Ltd.; S-propyl- L-cysteine sulfoxide (PCSO, CAS:17795-24-3, HPLC $\geq$ 98%) was purchased from Jiangsu Aikon Biopharmaceutical R&D Co., Ltd.; (E)-S-(1-propenyl)- L-cysteine sulfoxide (PeCSO) was identified by high-resolution mass spectrometry (MS). Twelve amino acids, asparagine (Asn), glutamine (Gln), serine (Ser), arginine (Arg), threonine (Thr), alanine (Ala), proline (Pro), methionine (Met), valine (Val), tryptophan (Trp), leucine (Leu), phenylalanine (Phe), HPLC $\geq$ 98%) were all purchased from Shanghai Yuanye Bio-Technology Co., Ltd.

2.3. Determination of Chlorophyll Content and Yellowing Rate

Chlorophyll content was measured as reported by [21] with slight modifications. Chive tissue (0.5 g) was ground with 5 mL of 95% ethanol (v/v) and then centrifuged at 12,000 g for 15 min at 4 °C. The chlorophyll content was quantified by measuring the absorbances at 665 and 649 nm in the supernatant and the content was expressed as g kg^{-1}. Except for the content of free amino acids, PeCSO and ACSOs, all the other results were expressed on a fresh weight basis.

At each time point, 1 kg chives from each sampling treatment were used to measure the yellowing rate. The leaf yellowing scales were divided into five grades: grade 0: no yellowing area; grade 1: approximately 1–10% yellowing area; grade 2: approximately 10–25% yellowing area; grade 3: approximately 25–50% yellowing area; and grade 4: more than 50% yellowing area. Leaf yellowing rate = $\sum$ (yellowing grade $\times$ number of leaves at this grade)/(total number of leaves $\times$ the highest grade).

2.4. Determination of Respiration Rate, MDA and H_2O_2

Respiration rate was directly measured by a CO_2 gas analyzer (CheckMate 3, Dansensor, Denmark). Approximately 400 g chives from each biological replicate were enclosed in 7.4 L glass jars at 20 °C or 3 °C for 1 h to measure respiration rate.

Hydrogen peroxide (H_2O_2) content and malondialdehyde (MDA) content were measured using the method of [22]. H_2O_2 content and MDA content were expressed as mmol kg^{-1}.

2.5. Determination of Antioxidants

ASA concentration was determined by the method as described by [23] and was expressed as g kg^{-1}. GSH (reduced glutathione) and GSSG (oxidized glutathione) concentrations were measured by an enzymatic cycling assay method described by [24] and were expressed as µmol kg^{-1}.

2.6. Determination of Antioxidant Enzymes

Superoxide dismutase (SOD) and ascorbic peroxidase (APX) activity measurements were adapted from [25]. One unit of SOD activity was defined as the amount of enzyme that caused 50% inhibition of nitro blue tetrazolium (NBT) at 560 nm. One unit of APX enzyme activity was defined as a decrease in absorbance by 0.001 per minute at 290 nm under the assay conditions.

Catalase (CAT) activity was determined according to the method reported by [26] with a slight modification. Chive tissue (0.5 g) was ground using 6 mL phosphate buffer (0.1 mol/L, pH 7.5, containing 5 mmol/L 1,4-Dithiothreitol and 5% polyvinyl pyrrolidone). The reaction mixture consisted of 2.8 mL sodium phosphate buffer (50 mmol/L, pH 7.5, containing 20 mmol/L H_2O_2) and 0.2 mL enzyme extract, with a total volume of 3.0 mL.

H_2O_2 decomposition was measured by the decrease in absorbance at 240 nm. One unit was defined as the amount of enzyme that caused one absorbance change per minute under the assay conditions.

Peroxidase (POD) activity was assayed as previously described by [27]. POD activity was measured based on guaiacol oxidation by H_2O_2 at 470 nm. The absorbance at 470 nm was recorded every 30 s using a spectrophotometer. One unit of enzyme activity was defined as the amount of enzyme required to increase the absorbance by 0.001 min^{-1} under the assay conditions.

2.7. Determination of GTP and Allinase

The γ-glutamyl transpeptidase (GTP) activity was determined according to the method reported by [28] with a slight modification. The 1 g samples were ground with 5 mL pre-cooled Tris-HCl (0.05 mol/L, containing 1 mol/L NaCl, 5 mmol/L 6-aminocaproic acids, 1 mmol/L phenylmethanesulfonyl fluoride) buffer at 4 °C, centrifuged at 12,000 g for 15 min at 4 °C. The reaction mixture, containing 0.2 mL enzyme extracts, 1 mL of 4 mmol/L γ-glutamyl p-nitroanilide and 1 mL of 40 mmol/L L-methionine, was incubated at 37 °C for 30 min. The absorbance was measured at 410 nm against a blank (enzyme extract substituted by distilled water). One unit of GTP activity is defined as the amount of enzyme that liberates 1 µmol of nitroaniline per min under standard conditions.

Allinase activity was determined according to the method reported by [29] with a slight modification. The 1 g samples were ground with 5 mL pre-cooled (4 °C) phosphate buffer (50 mmol/L, containing 5 mmol/L EDTA-Na_2, 25 µmol/L pyridoxal phosphate, pH 7.0), then centrifuged at 12,000 g for 15 min at 4 °C. The reaction mixture was 0.1 mL above enzyme extracts and 10 mmol/L S-methyl cysteine sulfoxide. The reaction mixture was incubated at 37 °C for 10 min. The enzymatic reaction was terminated by the addition of 1 mL of 2 mol/L HCl containing 5 mmol/L dinitrophenyl hydrazine. Subsequently, the mixture was incubated at 37 °C for 5 min, afterwards, 5 mL of 2.5 mol/L NaOH was added to the solution, incubated for 10 min and the absorbance was measured at 420 nm against a blank (enzyme extract substituted by distilled water). One unit of allinase activity is defined as the amount of enzyme that produces 1 µmol pyruvate per min under standard conditions.

2.8. Determination of Organosulfur Compounds and Amino Acids

The organosulfur compounds and amino acids were extracted and determined according to the method reported by [30] with a slight modification. An amount of 10 g chives were steamed with boiling water for 8 min, after being cooled, ground into a powder with a blender in liquid nitrogen and carefully transferred to a glass container. The sample powder was extracted with 100 mL 90% aqueous methanol containing 10 mmol/L HCl for 20 min with a magnetic stirrer at 40 °C, filtered and the residue was extracted with another 100 mL methanol again for 10 min as the same operation above. The combined methanolic extracts were concentrated at reduced pressure (at 40 °C) to approximately 10–15 mL and adjusted to 25 mL by 20 mmol/L borate buffer (pH 9.2). This extract was stored in the refrigerator (−20 °C) until derivatization.

Dansyl derivatives were prepared by mixing 100 µL of the sample extract with 250 µL of the dansyl chloride (Dns-Cl) reagent (10 mmol/L Dns-Cl in acetonitrile) and 0.65 mL of 20 mmol/L borate buffer (pH 9.2). The mixture was briefly shaken, allowed to stand at room temperature for 30 min, filtered through a 0.45 µm nylon filter, and analyzed by HPLC.

A high-performance liquid chromatography (HPLC) (LC-20AD, Shimadzu, Japan) equipped with a ZORBAX SB-Aq (Agilent, USA, 4.6 × 250 mm, 5 µm) column and PDA detector (SPD-M20A, Shimadzu, Japan) was used to determine the content of organosulfur compounds and amino acids. The chromatographic condition was as follows: 50 mmol/L sodium acetate buffer (pH 5.0, solvent A) and methanol (solvent B) were used as the mobile phase, with a flow rate of 0.9 mL min−1 and the gradient A/B 70/30 (0 min), 70/30 (2 min), 60/40 (in 12 min), 50/50 (in 32 min), 40/60 (in 42 min), 5/95 (in 42.01 min), 5/95 (in 51 min), 70/30 (in 51.01 min), and 70/30 (in 60 min), detection wavelength of 250 nm. The injection volume was 20 μL. The content of free amino acids, PeCSO and ACSOs, were expressed on a dry weight basis.

For identification of each peak, the eluents corresponding to each peak of HPLC were collected separately, desalted by C-18 solid-phase extraction column and analyzed by LC-MS system (G2-XS QTof, Waters). A 2 μL solution was injected into the UPLC column (2.1 × 100 mm ACQUITY UPLC BEH C18 column containing 1.7 μm particles) with a flow rate of 0.3 mL/min. Buffer A consisted of 0.1% formic acid in the water, and buffer B consisted of 0.1% formic acid in acetonitrile. The gradient was 5% buffer B for 0.5 min, 5–95% buffer B for 11 min, 95% buffer B for 2 min.

Mass spectrometry was performed using an electrospray source in positive ion mode with MS acquisition mode, with a selected mass range of 50–1200 m/z. The lock mass option was enabled using leucine-enkephalin (m/z 556.2771) for recalibration. The ionization parameters were the following: the capillary voltage was 2.5 kV, sample cone was 40 V, source temperature was 120 °C (and desolvation gas temperature was 400 °C). Data acquisition and processing were performed using Masslynx 4.1.

2.9. Statistical Analysis

All values are reported as the mean ± standard errors of three biological replicates. Statistical analysis was performed with the SPSS 18.0 software (SPSS Inc, Chicago, IL, USA). The significant differences between mean values were analyzed using the LSD's test ($p < 0.05$). Origin 2021 (Microcal Software, Northampton, MA, USA) was employed to create figures.

3. Results

3.1. Yellowing Rate, Chlorophyll Content and Appearance Analysis

The yellowing rate is a sensory quality while chlorophyll content is a physiological index used to assess the color status of chives. As shown in Figure 1, chives gradually lost the green color and turned yellow during storage. Leaves in the CK group were more yellow than 1-MCP treated chives on day 4 and day 5 during storage at 20 °C. In agreement with the appearance, the content of chlorophyll in all groups decreased constantly during storage, and the yellowing rate increased progressively. The chlorophyll content was 11.8% and 9.2% higher in 1-MCP treated samples on day 4 and day 5, respectively, than that in the CK group during RT storage. The yellowing rate of 1-MCP treated chives was 67.8% and 34.5% lower than that of CK samples on day 4 and day 5, respectively, during RT storage and negatively correlated with the chlorophyll content ($r = -0.89$ **). However, there was no difference in both the yellowing rate and the chlorophyll content of the 1-MCP treated and CK chives during storage at 3 °C. The result suggests that 1-MCP treatment could delay degradation of chlorophyll in chives, maintain green color and prolong the shelf life during RT storage. Nonetheless, the 1-MCP treatment did not improve the storage quality of the chives during LT storage.

Figure 1. Appearance (**A**), yellowing rate (**B**) and total chlorophyll content (**C**) of 1-MCP treated and untreated chives during storage at 20 °C and 3 °C. Data are expressed as the mean of triplicate samples. Vertical bars represent the standard errors of the means. The different letters at each time point indicate a difference among the chives samples.

3.2. Respiration, H_2O_2 and MDA Analysis

The respiration rate (Figure 2A) decreased sharply during the first 12 h of storage and then decreased at a slower rate afterward in all the experimental groups. Nevertheless, when stored at 20 °C, the respiration rate of 1-MCP treated chives was 13.7% and 17.0% lower than that of the CK group on day 2 and day 4. There was no difference between the respiration rate of the 1-MCP treated chives and the CK group stored during LT storage.

The H_2O_2 content in chives fluctuated, increasing at 12 h after harvest, dropping on day 2, and increased rapidly thereafter (Figure 2B). When stored at 20 °C, the H_2O_2 content in 1-MCP treated chives was 40.0% and 21.1% lower than that in the CK group on day 2 and day 5, respectively. However, when stored at 3 °C, the H_2O_2 content in the 1-MCP treated samples was higher than that in the CK group on day 2, with no difference afterward. This result suggested that 1-MCP treatment may have induced oxidative stress on chives transiently during LT storage.

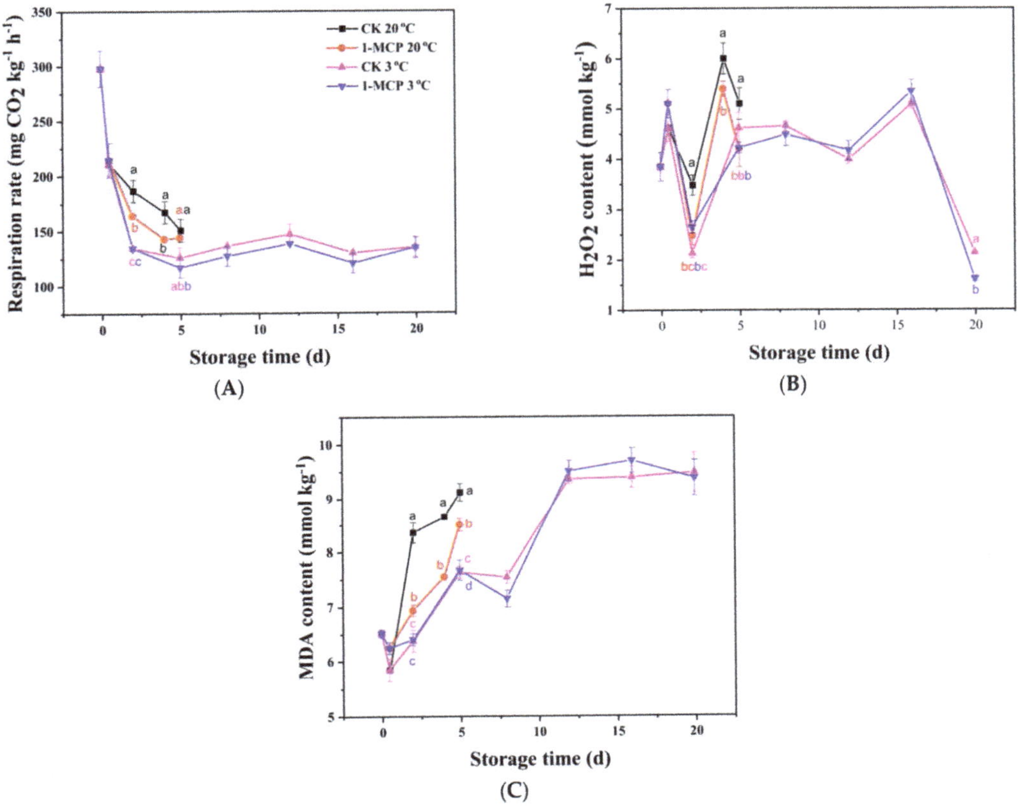

Figure 2. Respiration rate (**A**), H_2O_2 content (**B**) and MDA content (**C**) of 1-MCP treated and untreated chives during storage at 20 °C and 3 °C. Data are expressed as the mean of triplicate samples. Vertical bars represent the standard errors of the means. The different letters at each time point indicate a difference among the chives samples.

The MDA contents, which represent an index used to assess the extent of oxidative damage to the membrane, increased progressively in all the groups during storage (Figure 2C). Nonetheless, the 1-MCP treatment reduced the MDA content in the chives compared to the CK during RT storage. The MDA content was 20.5% and 14.7% lower in the 1-MCP sample than that in the CK group on days 2 and 4 of storage at RT, respectively. The MDA content in samples stored in 3 °C increased sharply after 8 days, reaching a maximum on day 12, and did not change after that.

The results clearly show that the 1-MCP treatment suppressed respiration rate, inhibited H_2O_2 and MDA accumulation, and protected the membrane from oxidative damage during RT storage, while the 1-MCP treatment had no protective effect on the quality characteristics of the chives during LT storage.

3.3. Non-Enzymatic Antioxidants Analysis

As shown in Figure 3A,B, GSH content in chives presented an overall decreasing trend except increasing on day 4 during RT storage, while GSH content declined during the first 5 days, and then increased progressively during storage at LT. GSSG content increased firstly and decreased afterward during storage at both temperatures. The 1-MCP increased GSH content by 10.0% and 18.3%, respectively, on day 4 and day 5, and suppressed GSSG content from day 0.5 to day 5, with 24.4% and 27.8% lower rates in 1-MCP treated chives on day 4 and day 5.

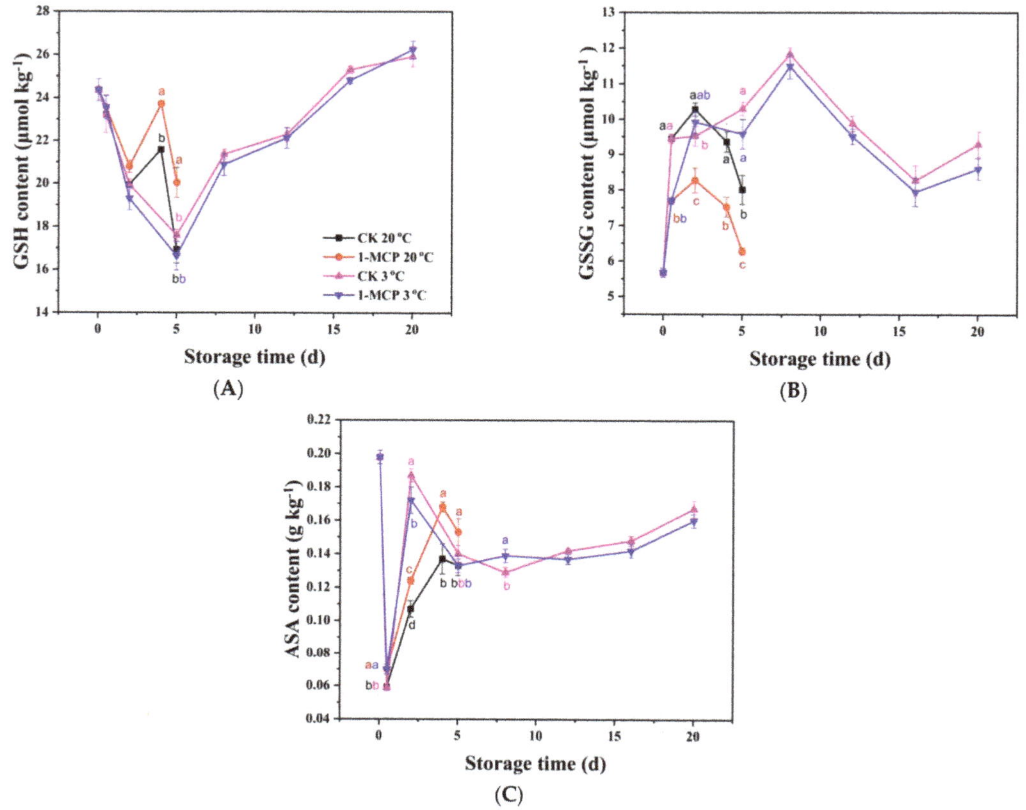

Figure 3. GSH content (**A**), GSSG content (**B**) and ASA content (**C**) of 1-MCP treated and untreated chives during storage at 20 °C and 3 °C. Data are expressed as the mean of triplicate samples. Vertical bars represent the standard errors of the means. The different letters at each time point indicate a difference among the chive samples.

With a negative correlation with GSSG (r = −0.71 **), ASA content (Figure 3C) in chives decreased sharply at the 12th hour, increased afterward and decreased again during storage. This might be due to the antioxidant activity of the ASA which culminated in reduce ROS accumulation, while the continuous synthesis of ASA might have been the reason for the fluctuation of ASA content. The 1-MCP treated chives had a higher ASA content during RT storage, 23.2% and 13.7% higher on day 4 and day 5 than untreated chives, respectively, while during LT storage, the 1-MCP treatment exerted a small effect on the ASA and GSH content of the chives.

3.4. Antioxidant Enzymes Analysis

Antioxidant enzymes in fruit and vegetables include SOD, CAT, APX, and POD. SOD is one of the most important antioxidant enzymes, and its activity declined in the first 12 h, increased afterward, dropped again during storage at RT. Nevertheless, SOD activity (Figure 4A) was 15.0%, 58.9% higher on day 2 and day 4, respectively, during RT storage in the 1-MCP treated sample than the CK. This result suggested that 1-MCP treatment could enhance SOD activity. The CAT activity (Figure 4B) and APX activity (Figure 4C) declined in the first 12 h after harvest in all groups, followed by a progressive increase. The CAT and APX showed a strong positive correlation (r = 0.84 **). When stored at 3 °C, the CAT and APX activity was unchanged from day 2 to day 8/12, after that it increased sharply at the late stage of storage in chives. The 1-MCP treatment improved the CAT activity in the treated sample during RT and LT storage. The 1-MCP treated chives had 29.4% and 28.3% higher CAT activity than untreated chives on day 2 and day 4, respectively, during RT storage, and 19.9% and 26.8% higher CAT activity than untreated chives on day 16 and day 20 during LT storage. The 1-MCP treatment also increased the APX activity during RT storage, with 9.2% and 9.4% higher APX activity in 1-MCP treated chives on day 4 and day 5. However, when stored at 3 °C, the APX activity in 1-MCP treated chives was higher than that in the CK group only on day 2; after that there was no difference in the APX activity. The results thus demonstrate that the 1-MCP treatment increased the activities of antioxidant enzymes in the chives during RT storage, but did not improve the antioxidant enzyme activities during LT storage.

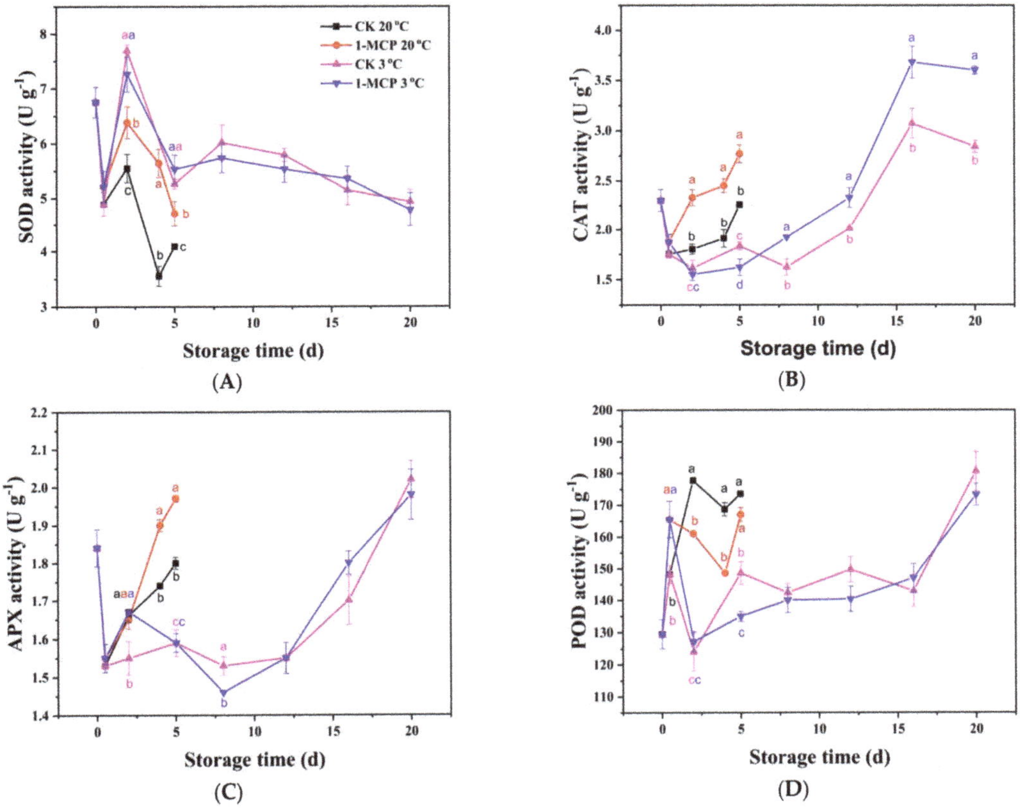

Figure 4. SOD activity (**A**), CAT activity (**B**), APX activity (**C**) and POD activity (**D**) of 1-MCP treated and untreated chives during storage at 20 °C and 3 °C. Data are expressed as the mean of triplicate samples. Vertical bars represent the standard errors of the means. The different letters at each time point indicate a difference among the chives samples.

There was no correlation between SOD/CAT/APX activity and POD activity. The POD activity increased dramatically in the first 12 h after harvest in all groups (Figure 4D). When stored at 20 °C, 1-MCP treatment inhibited the POD activity from day 2. When stored at 3 °C, the POD activity dropped on day 2, then increased from day 2 to day 16, with a more dramatic increase on day 20 for both the 1-MCP treated chives and the CK. The POD activity in the 1-MCP treated chives was higher on day 5 than that in CK, with no difference after that.

The results strongly demonstrate that 1-MCP treatment can efficiently enhance non-enzymatic and enzymatic antioxidant capacities in chives during RT storage, but do not affect the non-enzymatic and enzymatic antioxidant capacities of chives during LT storage.

3.5. Free Amino Acids Analysis

Based on the results of the preliminary experiment (1-MCP treatment had no effect on chives during LT storage), the content of ACSOs and free amino acids of 1-MCP treated chives stored at 3 °C were not determined. ACSOs and free amino acid content of the extract from chive samples are as shown in the HPLC chromatogram in Figure S1A. In postharvest vegetables, the degradation of protein, which is the result of vegetable senescence, will lead to an increase in the concentration of free amino acids. Generally, the concentration of almost all the identified free amino acids including Asn, Gln, Ser, Arg, Thr, Ala, Val, Trp, Leu, Phe in chive samples during storage at 20 °C and 3 °C was increased progressively, except Pro and Met (Figure S2). The content of total free amino acids increased at the 12th hour after harvest and further increased more dramatically at the end of storage, with a higher increase in their concentrations during RT storage, compared with LT storage (Figure 5). The Pro content in chives increased continuously when stored at RT but was not changed during storage at LT. A similar trend was observed in the Met content of the chives stored at 20 °C and 3 °C. Surprisingly, the 1-MCP treatment did not decrease the concentrations of total free amino acids during RT storage. The results thus show that the 1-MCP treatment could not inhibit the free amino acid content increasing.

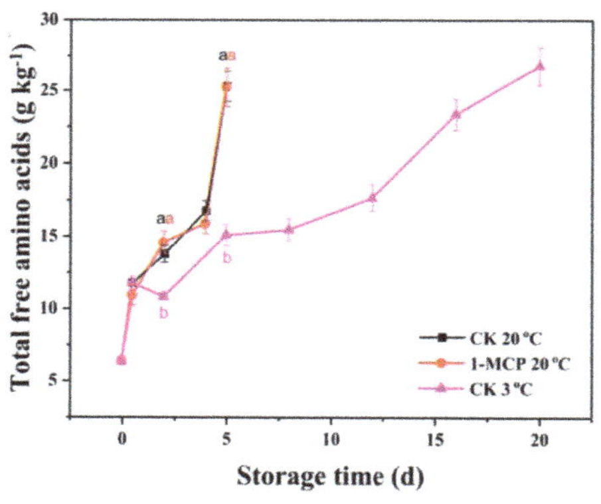

Figure 5. The content of total free amino acids of 1-MCP treated and untreated chives during storage at 20 °C and 3 °C. The content of free amino acids in 1-MCP treated chives stored at 3 °C was not determined. Data are expressed as the mean of triplicate samples. Vertical bars represent the standard errors of the means. The different letters of each treatment indicate a difference among the chives samples.

3.6. Organosulfur Compounds and Related Enzymes Analysis

Allinase, which is located in the cell vacuoles, hydrolyzes ACSOs, producing a variety of antibacterial sulfur-containing compounds [6]. The allinase activity increased dramatically in the first two days after harvest in all groups, then changed less dramatically after that (Figure 6A). When stored at 20 °C, the allinase activity in 1-MCP treated chives was 13.0% and 15.3% higher than that in untreated chives after 12 h and on day 2, after that, there was no difference in allinase activity. The allinase activity in samples stored at 3 °C remained unchanged from day 2 to day 8, then increased on day 12 and finally decreased to the end of the storage. In contrast to allinase activity ($r = -0.84$ **), the GTP activity decreased progressively during storage. The GTP activity in the sample stored at 20 °C was observed to decrease more rapidly than that in the samples stored at 3 °C (Figure 6B). Nonetheless, the GTP activity was higher in 1-MCP treated samples than that in untreated samples during RT storage; while the 1-MCP treatment did not increase the allinase activity and maintain GTP activity in chives during LT storage.

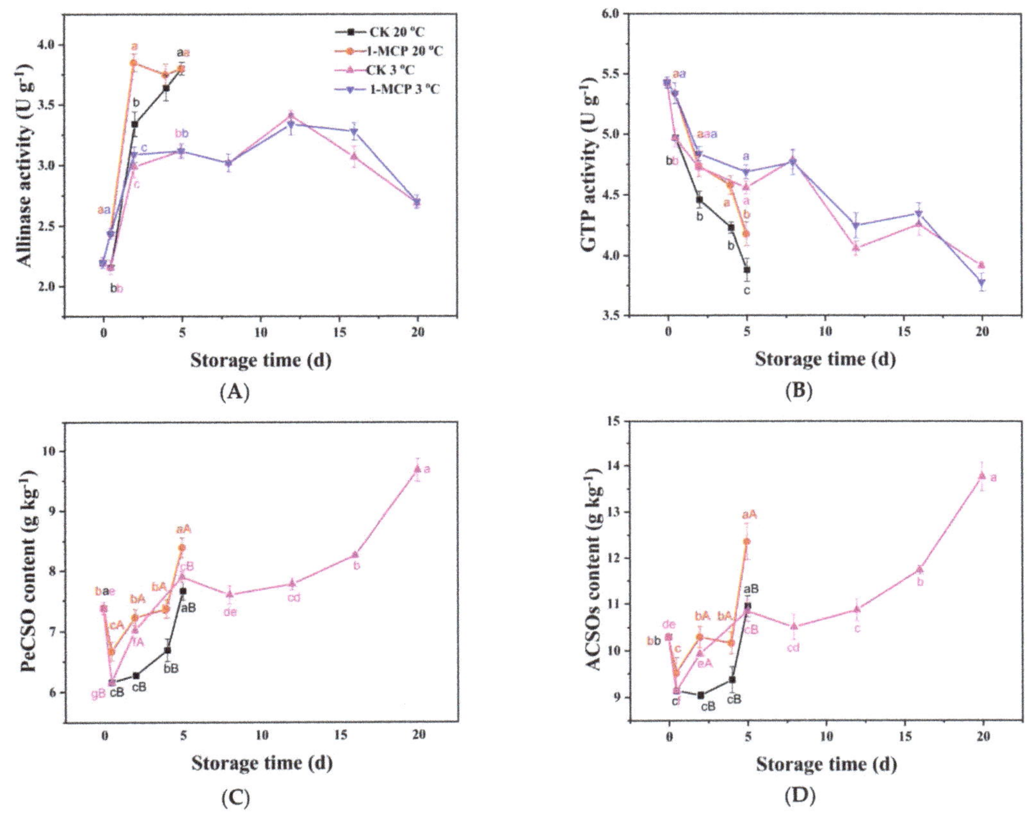

Figure 6. Allinase activity (**A**), GTP activity (**B**), PeCSO content (**C**) and ACSOs content (**D**) of 1-MCP treated and untreated chives during storage at 20 °C and 3 °C. The content of ACSOs in 1-MCP treated chives stored at 3 °C was not determined. Data are expressed as the mean of triplicate samples. Vertical bars represent the standard errors of the means. The lower and upper case letters indicate differences among the storage time and the treatments of the chive samples, respectively.

Isoalliin (PeCSO) was identified using LC-MS coupled with high-resolution mass spectrometry, as shown in Figure S1B,C. S-alk(en)ylcysteine sulfoxides (ACSOs) are bioactive organosulfur compounds in chives and include alliin (ACSO), isoalliin (PeCSO), methiin (MCSO), propiin (PCSO) [6]. The PeCSO is the predominant component of ACSOs in chives, and change in the PeCSO content was observed to show a similar trend with ACSOs content during storage (r = 0.96 **) (Figure 6C,D). MCSO was the second major component of ACSOs in the chives (Table 2), with the ratio of 3:1 of PeCSO to MCSO, while the concentration of ACSO and PCSO were very low in the chives. The PeCSO and ACSOs content in all groups declined after 12 h but increased at the end of storage. The PeCSO and ACSOs content in 1-MCP treated samples was higher compared to the untreated chives during RT storage, with a 15.3% and 13.9% increase, respectively, on day 2. The PeCSO and ACSOs content in samples stored at 3 °C increase from the 12th hour to day 5 with no further increase until day 12. The PeCSO and ACSOs then surged higher after that, with a 26.4% and 25.8% increase at the end of the storage than that in the CK group stored at 20 °C.

Table 2. S-alk(en)ylcysteine sulfoxides (ACSOs) content in chives.

Compounds Name	Storage Period (d)	0	0.5	2	4	5	8	12	16	20
MCSO (g kg^{-1})	CK 20 °C	2.44 ± 0.08 a	2.23 ± 0.00 b	2.10 ± 0.02 c,C	1.99 ± 0.09 d	2.16 ± 0.07 b,B	nd	nd	nd	nd
	1-MCP 20 °C	2.44 ± 0.08 b	2.12 ± 0.10 c	2.40 ± 0.08 b,A	2.08 ± 0.08 c	2.70 ± 0.12 a,A	nd	nd	nd	nd
	CK 3 °C	2.44 ± 0.08 b	2.23 ± 0.00 c,d	2.25 ± 0.01 c,B	nd	2.12 ± 0.14 d,e,B	2.04 ± 0.10 e	2.15 ± 0.13 c,d,e	2.52 ± 0.02 b	2.85 ± 0.11 a
ACSO (g kg^{-1})	CK 20 °C	0.35 ± 0.01 c	0.62 ± 0.06 b,A	0.57 ± 0.04 b	0.57 ± 0.06 b	1.01 ± 0.04 a,A	nd	nd	nd	nd
	1-MCP 20 °C	0.35 ± 0.01 c	0.41 ± 0.04 c,B	0.56 ± 0.03 b	0.62 ± 0.05 b	1.06 ± 0.09 a,A	nd	nd	nd	nd
	CK 3 °C	0.35 ± 0.01 e	0.62 ± 0.06 d,A	0.56 ± 0.03 d	nd	0.72 ± 0.05 c,B	0.75 ± 0.02 c	0.82 ± 0.06 b	0.83 ± 0.07 b	1.07 ± 0.09 a
PCSO (g kg^{-1})	CK 20 °C	0.13 ± 0.00 a	0.12 ± 0.00 b,B	0.11 ± 0.00 d,B	0.11 ± 0.01 d,A	0.12 ± 0.00 c	nd	nd	nd	nd
	1-MCP 20 °C	0.13 ± 0.00 b	0.23 ± 0.01 a,A	0.12 ± 0.01 c,A	0.09 ± 0.01 d,B	0.12 ± 0.01 b,c	nd	nd	nd	nd
	CK 3 °C	0.13 ± 0.00 b	0.12 ± 0.00 c,B	0.10 ± 0.00 d,B	nd	0.12 ± 0.00 c	0.09 ± 0.01 e	0.12 ± 0.00 c	0.13 ± 0.00 b	0.16 ± 0.01 a
PeCSO (g kg^{-1})	CK 20 °C	7.38 ± 0.17 a	6.16 ± 0.01 c,B	6.27 ± 0.06 c,B	6.69 ± 0.32 b,B	7.66 ± 0.26 a,B	nd	nd	nd	nd
	1-MCP 20 °C	7.38 ± 0.17 b	6.67 ± 0.27 c,A	7.23 ± 0.22 b,A	7.38 ± 0.25 b,A	8.39 ± 0.29 a,A	nd	nd	nd	nd
	CK 3 °C	7.38 ± 0.17 e	6.16 ± 0.01 g,B	7.02 ± 0.04 f,A	nd	7.90 ± 0.16 c,B	7.60 ± 0.26 d,e	7.78 ± 0.17 c,d	8.26 ± 0.07 b	9.68 ± 0.33 a
Total ACSOs (g kg^{-1})	CK 20 °C	10.30 ± 0.03 b	9.14 ± 0.07 c	9.04 ± 0.12 c,B	9.37 ± 0.48 c,B	10.95 ± 0.37 a,B	nd	nd	nd	nd
	1-MCP 20 °C	10.30 ± 0.03 b	9.43 ± 0.58 c	10.30 ± 0.39 b,A	10.17 ± 0.42 b,A	12.27 ± 0.69 a,A	nd	nd	nd	nd
	CK 3 °C	10.30 ± 0.03 d,e	9.14 ± 0.07 f	9.93 ± 0.07 e,A	nd	10.85 ± 0.38 c,B	10.48 ± 0.47 c,d	10.87 ± 0.40 c	11.74 ± 0.16 b	13.77 ± 0.53 a

Note: nd means not detect. Total ACSOs content is the sum of the content of MCSO, ACSO, PeCSO, and PCSO. Data are expressed as the mean of triplicate samples ± standard errors. The lower and upper case letters indicate differences among the storage time and the treatments of the chive samples, respectively.

4. Discussion

4.1. 1-MCP Enhances Antioxidant Capacity and Delays Senescence in Postharvest Chives

Through preliminary experiments, we found that chives could only be stored for 5 days at 20 °C and only 20 days at 3 °C. Thus we chose 5 days of study for 20 °C and 20 days of study for 3 °C. Vegetables do not require ethylene action for normal senescence to take place [31,32]. In this study, the ethylene concentration determined in chive was very low, with no significant difference between 1-MCP treatment and CK group under the condition by using Agilent GC system (6890N) and Agilent J&W GC columns (19095P-S23). Therefore, we did not show the result of ethylene production of chives.

Respiration is an indication of the tissue metabolic rate and the major factor that contributes to the deterioration and loss of quality of plant produce during postharvest storage [33]. Chive leaves, with a very high respiratory rate, suffer senescence very quickly after harvest. The 1-MCP treatment suppressed the respiration rate during RT storage, which is consistent with other reports [13,34]. Respiration, a basic biological process in a plant cell, produces reactive oxygen species (ROS) as an unavoidable by-product [8]. The 1-MCP treatment reduced the respiration rate and its accompanying ROS accumulation in the chives during storage at RT. This was evident in the lower H_2O_2 content in 1-MCP treated chives during RT storage. However, during LT storage, 1-MCP treated chives showed no difference in respiration rate and H_2O_2 content compared with untreated chives. This result strongly indicates that 1-MCP treatment does not affect the physical activity and enzyme activity of chive when stored at LT. Similar results have been reported in a combined 1-MCP and heat treatment on peach [34].

The MDA concentration and ROS level are often used as an index to measure membrane integrity and oxidative stress in vegetables [34]. In this study, the MDA concentration in 1-MCP treated samples was lower than that in untreated samples during RT storage. It is also found that the lower concentrations of H_2O_2 in 1-MCP treated vegetables correlated with the lower MDA level during RT storage. This finding suggests that 1-MCP treatment effectively maintained the membrane integrity of the chive by suppressing oxidative stress during RT storage, while the 1-MCP treatment did not affect the MDA and ROS levels of the chives during LT storage in this study. It has been reported that combined 1-MCP and heat-treatment even induced oxidative stress during LT storage in peach [34].

The 1-MCP preserved a high content of antioxidants (ASA and GSH) and high activities of antioxidant enzymes (SOD, CAT, and APX) in chives stored at 20 °C which might have contributed to scavenging ROS (H_2O_2), and maintained the cell structural integrity. The accumulation of ROS in plants during senescing can activate the enzymatic and non-enzymatic antioxidant system to scavenge ROS down to a safe range [35]. These results showed that 1-MCP treatment enhanced the antioxidant system during RT storage, thus, reduced ROS production. This postulation was validated by the fact that the H_2O_2 content and MDA content was lower in 1-MCP treated chives during RT storage. The 1-MCP treatment had little effect on the antioxidant system during LT storage, except for CAT activity. It has been reported in the literature that POD activity increases as the leaves are senescing, the role of which is dual [36]. On one hand, as a member of the protective enzyme system, it can clear reactive oxygen species; while on the other hand it participates in the degradation of chlorophyll and the production of reactive oxygen species, and triggers membrane lipid peroxidation, manifested as a damaging effect. In this study, 1-MCP treatment-induced POD activity in the first 12 h and inhibited it afterward during RT storage. Similar reports have found that hyperbaric pressures between 400 kPa and 800 kPa induced a reduction in POD activity due to the reduction in oxidative stress and delayed senescence in tomatoes [37]. Vacuum packing at pressures of down arrow 0.04 mPa at 4 °C has also been reported to suppress POD activity and maintained the quality of mung bean [38].

Yellowing, the main characteristic of leaf senescence, with a strong positive correlation with MDA (r = 0.89 **), is the result of chlorophyll degradation. The 1-MCP treatment delayed chlorophyll degradation in chives during RT storage. Similar findings have been reported [15,39]. The protective effects of 1-MCP on chlorophyll degradation may be attributed to the higher antioxidant ability and lower ROS level, maintaining membrane integrity [39]. Higher chlorophyll content in the 1-MCP treated chives correlated with lower H_2O_2 content on days 4 and 5 than that in the CK group during RT storage. The 1-MCP treatment on the chives during LT storage did not prevent chlorophyll degradation in this study. This might be due to the inability of the 1-MCP treatment to decrease oxidative stress and maintain membrane integrity of the chive during LT storage.

When plant cells undergo senescence, they suffer from oxidative stress and produce a large amount of ROS, causing membrane lipid peroxidation, degradation of protein, and production of free amino acids [40,41]. The total free amino acids increased progressively at first, then increased more dramatically the last day (RT storage) and on the last 8 days during LT storage (Figure 5), which was an indication that the chloroplast had been severely degraded. This was also evident in the strong positive correlation between the content of free amino acids and MDA (r > 0.7 **), the content of free amino acids and yellowing rate (r > 0.8 **), and the strong negative correlation between free amino acids and chlorophyll content (r > −0.8 **). A similar result has been reported on the increase in free amino acids in mushrooms during postharvest storage [42]. In this study, the 1-MCP treatment did not affect the accumulation of free amino acids in the chives during storage which is similar to the report that both ice and the novel phase change material treatment to *Pleurotus eryngii* showed no effects on total free amino acids [43].

In summary, the results indicate that 1-MCP enhances antioxidant capacity, reduces oxidative stress, and delays senescence in postharvest chives during storage at RT, while 1-MCP treatment had no effect on the antioxidant capacity, oxidative stress, and senescence levels on chives during storage at LT.

4.2. ACSOs Probably Act as Antioxidants Being Increased by 1-MCP in Postharvest Chives

S-alk(en)ylcysteine sulfoxides (ACSOs) are bioactive compounds that are beneficial for both the plant itself and human health due to their antibiotic properties. It has been widely reported that alliin is an effective hydroxyl (OH) scavenger [44–46], it efficaciously scavenged ·OH more than ascorbic acid [45], and scavenged superoxide generated by the xanthine/xanthine oxidase system [44]. It has been reported that the content of pyruvic acid, the by-product of ACSOs enzymatically hydrolyzed by allinase, was strongly correlated with increased FRAP (ferric reducing antioxidant power) activity [47]. Some researchers have also reported garlic had the highest content of total organosulfur compounds (mainly alliin) during the 8 weeks of storage which positively correlated with the maximum antioxidant capacities exhibited at 8 weeks of the storage [48]. In this study, the content of PeCSO and ACSOs showed a strong positive correlation with CAT activity (r = 0.86 **, 0.80 **) and with APX activity (r = 76 **, 0.72 **), respectively, indicating that ACSOs in chives probably play a key role in antioxidant capacities during storage. ACSOs in chives increased during storage, and 1-MCP treatment induced an increase in the ACSOs content during RT storage. LT storage induced an increase in ACSOs, with higher ACSOs in chives stored at 3 °C than that in 20 °C at the end of storage. Similar studies have also observed an increase in the total amount of the major ACSOs (e.g., alliin, methiin, and isoalliin) of up to 30% when garlic bulbs of 58 genotypes were stored at 5 °C [49]. It has been found that after 5 months of cold storage, ACSOs and enzymatically produced pyruvic acid concentrations increased in 9 of 10 genotypes of onion [20]. Some reports [50] have suggested that the increase in pungency during storage is due to the release of (1-propenyl)-L-cysteine sulfoxide (isoalliin) from γ-glutamylcysteine sulfoxide by γ-glutamyl transpeptidase (GTP), which increases its activity up to five-fold just before the onion starts sprouting, while in this study, GTP activity declined progressively during storage in chives. The increase in ACSOs content with a decrease in GTP activity suggests that perhaps enzymes possibly exist which could hydrolyze γ-glutamylcysteine sulfoxide to produce ACSOs. It also further suggests that the ACSOs could have been biosynthesized from other compounds such as amino acids. [51]. There are two proposed routes for alliin biosynthesis, one is from glutathione via γ-glutamyl peptides, while the other is from serine and allyl thiol [6]. In this study, two sulfur amino acids, cysteine of a very low concentration, and methionine, which its concentration decreased during storage while increasing at the end of storage, may be involved in the biosynthesis of ACSOs. This further suggests that the GTP might not have been involved in the biosynthesis of ACSOs in the chive samples. Allinase is a protective enzyme that hydrolyzes ACSOs to produce bioactive compounds when the tissue is attacked by insects or a pathogen. Its activity increased sharply from day 2 of the storage and maintained high activity throughout the storage period under RT. However its activity declined on day 20 during storage at low temperature. This result suggested that allinase activity decreased at the end of LT storage as its physical activity declined. The normalized transcription of allinase enzymes has been reported to be very high during sprouting but declined at the senescent stage of leaves when the alliin had transferred to the bulb [52]. This indicates that the allinase activity can be induced by biotic or abiotic stress for self-protection and survival in plants. The 1-MCP treatment induced higher allinase activity in chives during RT storage, while it had little effect on chives' allinase activity during LT storage.

Overall, ACSOs showed a strong positive correlation with the CAT and APX, indicating that ACSOs may act as antioxidants in chives. The 1-MCP increased the content of ACSOs during storage at RT. However, compared with RT storage, chives stored at LT storage had higher ACSOs content.

5. Conclusions

The possible senescence mechanism of chive leaves and mechanism of yellowing inhibition of chives during storage by 1-MCP treatment is shown in Figure 7. The 100 μL/L 1-MCP treatment was effective in inhibiting ROS production, enhancing antioxidant ability, reducing membrane damage, delaying chive yellowing and senescence during RT storage. Compared with RT storage, 1-MCP treatment had no positive effect on delaying chive senescence during LT storage. However, the 1-MCP treatment (under RT) and LT storage (compared with RT storage) promoted the increase of ACSOs content in the chives during storage. The present study indicates that ACSOs in chives probably play a key role in antioxidant activity due to their strong positive correlation with the increase in antioxidant enzyme activities during storage. To the best of our knowledge, this is the first report on the correlation between ACSOs and antioxidant ability. In conclusion, 1-MCP treatment is more effective in improving chives quality and delaying senescence for chives stored at RT, compared to chives stored at LT. Thus the study efficiently demonstrates that 1-methylcyclopropene preserves the quality of chive (*Allium schoenoprasum* L.) by enhancing its antioxidant capacities and organosulfur profile during storage.

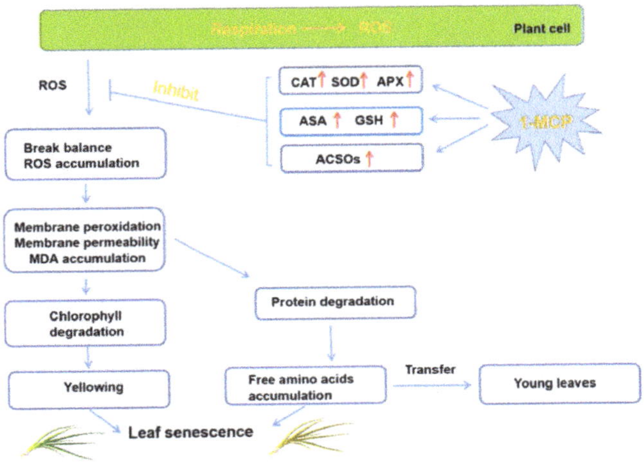

Figure 7. Possible senescence mechanism of chive leaves and possible mechanism of yellowing inhibition of chives during storage by 1-MCP treatment.

Supplementary Materials: The following are available online at https://www.mdpi.com/article/10.3390/foods10081792/s1, Figure S1: Separation and identification of organosulfur compounds and amino acids. HPLC chromatograms of organosulfur compounds and amino acids in chives (A), the peaks as the following: 1.Asn, 2. Gln, 3. Ser, 4. MCSO, 5. Arg, 6. Thr, 7. Ala, 8. PeCSO, 9. ACSO, 10. PCSO, 11. Pro, 12. Met, 13. Val, 14. Trp, 15. Leu, 16. Phe. TIC (B) and Mass spectrum (C) ($[M+H]^+$) of LC-MS chromatograms of separated peak 8 (a derivative of isoalliin with Dns-Cl) of chives extracts., Figure S2: Asn content (A), Gln content (B), Ser content (C), Arg content (D), Thr content (E), Ala content (F), Val content (G), Trp content (H), Leu content (I), Phe content (J), Pro content (K) and Met content (L) in 1-MCP treated chives and untreated chives during storage at 20 °C and 3 °C. The content of free amino acids in 1-MCP treated chives stored at 3 °C were not determined. Data are expressed as the mean of triplicate samples. Vertical bars represent the standard errors of the means. The lower and upper case letters indicate difference among the storage time and the treatments of the chive samples respectively.

Author Contributions: X.D.: conceptualization, methodology, software, writing an original draft. Y.L.: methodology, writing—review & editing. Y.Y.: investigation, validation, software. Z.Y.: supervision, project administration. Declaration of competing interest: the authors declare that we do not have any commercial or associative interest that represents a conflict of interest in connection with the work submitted. All authors have read and agreed to the published version of the manuscript.

Funding: This research was funded by the Key Project of Jiangsu Modern Agriculture (BE2018382).

Institutional Review Board Statement: Not applicable.

Informed Consent Statement: Not applicable.

Data Availability Statement: Not applicable.

Acknowledgments: The authors thank the College of Life Sciences (Nanjing Agricultural University) for providing support, and the LCMS-QTOF instrument for the identification of organosulfur compounds and amino acids in this study.

Conflicts of Interest: The authors declare no conflict of interest.

References

1. Chen, Z.; Ren, X.; Meng, X.; Zhang, Y.; Chen, D.; Tang, F. Novel Fluorescence Method for Detection of alpha-L-Fucosidase Based on CdTe Quantum Dots. *Anal. Chem.* **2012**, *84*, 4077–4082. [CrossRef]
2. Fredotović, Ž.; Soldo, B.; Šprung, M.; Marijanović, Z.; Jerković, I.; Puizina, J. Comparison of Organosulfur and Amino Acid Composition between Triploid Onion Allium cornutum Clementi ex Visiani, 1842, and Common Onion Allium cepa L., and Evidences for Antiproliferative Activity of Their Extracts. *Plants* **2020**, *9*, 98. [CrossRef] [PubMed]
3. Hirschegger, P.; Jakše, J.; Trontelj, P.; Bohane, B. Origins of Allium ampeloprasum horticultural groups and a molecular phylogeny of the section Allium (Allium: Alliaceae). *Mol. Phylogenet. Evol.* **2010**, *54*, 488–497. [CrossRef] [PubMed]
4. Kamenetsky, R.; Rabinowitch, H.D. Physiology of Domesticated Alliums: Onions, Garlic, Leek, and Minor Crops. In *Encyclopedia of Applied Plant Sciences*, 2nd ed.; Thomas, B., Murray, B.G., Murphy, D.J., Eds.; Academic Press: Oxford, UK, 2017; pp. 255–261.
5. Chen, H. Chives. In *Handbook of Herbs and Spices*; Woodhead Publishing: Sawston, UK, 2006; pp. 337–346.
6. Yoshimoto, N.; Saito, K. S-Alk(en)ylcysteine sulfoxides in the genus Allium: Proposed biosynthesis, chemical conversion, and bioactivities. *J. Exp. Bot.* **2019**, *70*, 4123–4137. [CrossRef] [PubMed]
7. Ougham, H.J.; Morris, P.; Thomas, H. The colors of autumn leaves as symptoms of cellular recycling and defenses against environmental stresses. *Curr. Top. Dev. Biol.* **2005**, *66*, 135–160. [CrossRef] [PubMed]
8. Suzuki, N.; Miller, G.; Morales, J.; Shulaev, V.; Angel Torres, M.; Mittler, R. Respiratory burst oxidases: The engines of ROS signaling. *Curr. Opin. Plant Biol.* **2011**, *14*, 691–699. [CrossRef]
9. Tian, S.; Qin, G.; Li, B. Reactive oxygen species involved in regulating fruit senescence and fungal pathogenicity. *Plant Mol. Biol.* **2013**, *82*, 593–602. [CrossRef]
10. Li, F.; Huang, H.; Ding, X.; Liu, J.; He, M.; Shan, Y.; Qu, H.; Jiang, Y. Effect of CPPU on postharvest attributes of Chinese flowering cabbage during storage. *Postharvest Biol. Technol.* **2021**, *174*, 111438. [CrossRef]
11. Sun, B.; Yan, H.; Liu, N.; Wei, J.; Wang, Q. Effect of 1-MCP treatment on postharvest quality characters, antioxidants and glucosinolates of Chinese kale. *Food Chem.* **2012**, *131*, 519–526. [CrossRef]
12. Xu, F.; Wang, H.; Tang, Y.; Dong, S.; Qiao, X.; Chen, X.; Zheng, Y. Effect of 1-methylcyclopropene on senescence and sugar metabolism in harvested broccoli florets. *Postharvest Biol. Technol.* **2016**, *116*, 45–49. [CrossRef]
13. Al Ubeed, H.M.S.; Wills, R.B.H.; Bowyer, M.C.; Golding, J.B. Comparison of hydrogen sulphide with 1-methylcyclopropene (1-MCP) to inhibit senescence of the leafy vegetable, pak choy. *Postharvest Biol. Technol.* **2018**, *137*, 129–133. [CrossRef]

14. Jiang, L.; Kang, R.; Zhang, L.; Jiang, J.; Yu, Z. Differential protein profiles of postharvest Gynura bicolor D.C leaf treated by 1-methylcyclopropene and ethephon. *Food Chem.* **2015**, *176*, 27–39. [CrossRef] [PubMed]
15. Hassan, F.A.S.; Mahfouz, S.A. Effect of 1-methylcyclopropene (1-MCP) on the postharvest senescence of coriander leaves during storage and its relation to antioxidant enzyme activity. *Sci. Hortic.* **2012**, *141*, 69–75. [CrossRef]
16. Akan, S.; Tuna Gunes, N.; Yanmaz, R. Methyl jasmonate and low temperature can help for keeping some physicochemical quality parameters in garlic (Allium sativum L.) cloves. *Food Chem.* **2019**, *270*, 546–553. [CrossRef]
17. Chope, G.A.; Cools, K.; Hammond, J.P.; Thompson, A.J.; Terry, L.A. Physiological, biochemical and transcriptional analysis of onion bulbs during storage. *Ann. Bot.* **2012**, *109*, 819–831. [CrossRef] [PubMed]
18. Kamata, Y.; Aoyagi, M.; Sawada, Y.; Nakabayashi, R.; Hirai, M.Y.; Saito, K.; Imai, S. Changes in trans-S-1-Propenyl-L-cysteine Sulfoxide and Related Sulfur-Containing Amino Acids during Onion Storage. *J. Agric. Food Chem.* **2016**, *64*, 9063–9071. [CrossRef]
19. Naheed, Z.; Cheng, Z.; Wu, C.; Wen, Y.; Ding, H. Total polyphenols, total flavonoids, allicin and antioxidant capacities in garlic scape cultivars during controlled atmosphere storage. *Postharvest Biol. Technol.* **2017**, *131*, 39–45. [CrossRef]
20. Romo-Pérez, M.L.; Weinert, C.H.; Häußler, M.; Egert, B.; Frechen, M.A.; Trierweiler, B.; Kulling, S.E.; Zörb, C. Metabolite profiling of onion landraces and the cold storage effect. *Plant Physiol. Biochem.* **2020**, *146*, 428–437. [CrossRef] [PubMed]
21. Lande, N.V.; Subba, P.; Barua, P.; Gayen, D.; Prasad, T.S.K.; Chakraborty, S.; Chakraborty, N. Dissecting the chloroplast proteome of chickpea (Cicer arietinum L.) provides new insights into classical and non-classical functions. *J. Proteom.* **2017**, *165*, 11–20. [CrossRef]
22. Du, Q.; Zhao, X.-H.; Xia, L.; Jiang, C.-J.; Wang, X.-G.; Han, Y.; Wang, J.; Yu, H.-Q. Effects of potassium deficiency on photosynthesis, chloroplast ultrastructure, ROS, and antioxidant activities in maize (Zea mays L.). *J. Integr. Agric.* **2019**, *18*, 395–406. [CrossRef]
23. Lin, Y.; Lin, H.; Zhang, S.; Chen, Y.; Chen, M.; Lin, Y. The role of active oxygen metabolism in hydrogen peroxide-induced pericarp browning of harvested longan fruit. *Postharvest Biol. Technol.* **2014**, *96*, 42–48. [CrossRef]
24. Griffith, O.W. Determination of glutathione and glutathione disulfide using glutathione reductase and 2-vinylpyridine. *Anal. Biochem.* **1980**, *106*, 207–212. [CrossRef]
25. Han, C.; Li, J.; Jin, P.; Li, X.; Wang, L.; Zheng, Y. The effect of temperature on phenolic content in wounded carrots. *Food Chem.* **2017**, *215*, 116–123. [CrossRef]
26. Tareen, M.J.; Abbasi, N.A.; Hafiz, I.A. Postharvest application of salicylic acid enhanced antioxidant enzyme activity and maintained quality of peach cv. 'Flordaking' fruit during storage. *Sci. Hortic.* **2012**, *142*, 221–228. [CrossRef]
27. Razavi, F.; Hajilou, J. Enhancement of postharvest nutritional quality and antioxidant capacity of peach fruits by preharvest oxalic acid treatment. *Sci. Hortic.* **2016**, *200*, 95–101. [CrossRef]
28. Zhao, F.; Qiao, X. Studies on the Purification and Partial Enzymology Characterization of γ-Glutamyl Transpeptidase in Garlic. *J. Chin. Inst. Food Sci. Technol.* **2009**, *9*, 41–45. [CrossRef]
29. Hanum, T.; Sinha, N.K.; Cash, J.N. Characteristics of gamma-glutamyl-transpeptidase and alliinase of onion and their effects on the enhancement of pyruvate formation in onion macerates. *J. Food Biochem.* **1995**, *19*, 51–65. [CrossRef]
30. Kubec, R.; Dadakova, E. Chromatographic methods for determination of S-substituted cysteine derivatives—A comparative study. *J. Chromatogr. A* **2009**, *1216*, 6957–6963. [CrossRef]
31. Botton, A.; Tonutti, P.; Ruperti, B. Biology and Biochemistry of Ethylene. In *Postharvest Physiology and Biochemistry of Fruits and Vegetables*; Woodhead Publishing: Sawston, UK, 2019; pp. 93–112.
32. Seymour, G.B.; Chapman, N.H.; Chew, B.L.; Rose, J.K.C. Regulation of ripening and opportunities for control in tomato and other fruits. *Plant Biotechnol. J.* **2013**, *11*, 269–278. [CrossRef] [PubMed]
33. Sohail, M.; Wills, R.B.H.; Bowyer, M.C.; Pristijono, P. Beneficial impact of exogenous arginine, cysteine and methionine on postharvest senescence of broccoli. *Food Chem.* **2021**, *338*, 128055. [CrossRef]
34. Huan, C.; An, X.; Yu, M.; Jiang, L.; Ma, R.; Tu, M.; Yu, Z. Effect of combined heat and 1-MCP treatment on the quality and antioxidant level of peach fruit during storage. *Postharvest Biol. Technol.* **2018**, *145*, 193–202. [CrossRef]
35. Liu, W.; Zheng, C.; Chen, J.; Qiu, J.; Huang, Z.; Wang, Q.; Ye, Y. Cold acclimation improves photosynthesis by regulating the ascorbate-glutathione cycle in chloroplasts of Kandelia obovata. *J. For. Res.* **2019**, *30*, 755–765. [CrossRef]
36. Yu, Z.; Zhang, W.; Yue, S.; Shen, C.; Yu, S. Effects of potassium nutrition on photosynthesis and senescence of winter wheat. *Acta Agron. Sin.* **1996**, 305–312.
37. Inestroza-Lizardo, C.; Mattiuz, B.-H.; da Silva, J.P.; Voigt, V.; Muniz, A.C.; Pinsetta, J.S. Effect of hyperbaric pressure on the activity of antioxidant enzymes and bioactive compounds of cv. 'Débora' tomatoes. *Sci. Hortic.* **2019**, *249*, 340–346. [CrossRef]
38. Zhang, S.J.; Hu, T.T.; Liu, H.K.; Chen, Y.Y.; Pang, X.J.; Zheng, L.L.; Chang, S.M.; Kang, Y.F. Moderate vacuum packing and low temperature effects on qualities of harvested mung bean (Vigna radiata L.) sprouts. *Postharvest Biol. Technol.* **2018**, *145*, 83–92. [CrossRef]
39. Song, L.; Yi, R.; Luo, H.; Jiang, L.; Gu, S.; Yu, Z. Postharvest 1-methylcyclopropene application delays leaf yellowing of pak choi (Brassica rapa subsp. chinensis) by improving chloroplast antioxidant capacity and maintaining chloroplast structural integrity during storage at 20 °C. *Sci. Hortic.* **2020**, *270*. [CrossRef]
40. Palma, J.M.; Sandalio, L.M.; Corpas, F.J.; Romero-Puertas, M.C.; McCarthy, I.; del Rio, L.A. Plant proteases, protein degradation, and oxidative stress: Role of peroxisomes. *Plant Physiol. Biochem.* **2002**, *40*, 521–530. [CrossRef]
41. Solomon, M.; Belenghi, B.; Delledonne, M.; Menachem, E.; Levine, A. The involvement of cysteine proteases and protease inhibitor genes in the regulation of programmed cell death in plants. *Plant Cell* **1999**, *11*, 431–443. [CrossRef] [PubMed]

42. Zhang, K.; Pu, Y.-Y.; Sun, D.-W. Recent advances in quality preservation of postharvest mushrooms (Agaricus bisporus): A review. *Trends Food Sci. Technol.* **2018**, *78*, 72–82. [CrossRef]
43. Li, D.; Wang, D.; Fang, Y.; Belwal, T.; Li, L.; Lin, X.; Xu, Y.; Chen, H.; Zhu, M.; Luo, Z. Involvement of energy metabolism and amino acid metabolism in quality attributes of postharvest Pleurotus eryngii treated with a novel phase change material. *Postharvest Biol. Technol.* **2021**, *173*. [CrossRef]
44. Chung, L.Y. The Antioxidant Properties of Garlic Compounds: Allyl Cysteine, Alliin, Allicin, and Allyl Disulfide. *J Med. Food* **2006**, *9*, 205–213. [CrossRef]
45. Kim, J.-M.; Chang, H.J.; Kim, W.-K.; Chang, N.; Chun, H.S. Structure−Activity Relationship of Neuroprotective and Reactive Oxygen Species Scavenging Activities for Allium Organosulfur Compounds. *J. Agric. Food Chem.* **2006**, *54*, 6547–6553. [CrossRef]
46. Kourounakis, P.N.; Rekka, E.A. Effect on active oxygen species of alliin and allium sativum (garlic) powder. *Res. Commun. Chem. Pathol. Pharmacol.* **1991**, *74*, 249–252.
47. Avgeri, I.; Zeliou, K.; Petropoulos, S.A.; Bebeli, P.J.; Papasotiropoulos, V.; Lamari, F.N. Variability in Bulb Organosulfur Compounds, Sugars, Phenolics, and Pyruvate among Greek Garlic Genotypes: Association with Antioxidant Properties. *Antioxidants* **2020**, *9*, 967. [CrossRef]
48. Fei, M.L.I.; Tong, L.I.; Wei, L.I.; De Yang, L. Changes in antioxidant capacity, levels of soluble sugar, total polyphenol, organosulfur compound and constituents in garlic clove during storage. *Ind. Crop. Prod.* **2015**, *69*, 137–142. [CrossRef]
49. Hornickova, J.; Kubec, R.; Cejpek, K.; Velisek, J.; Ovesna, J.; Stavelikova, H. Profiles of S-Alk(en)ylcysteine Sulfoxides in Various Garlic Genotypes. *Czech J. Food Sci.* **2010**, *28*, 298–308. [CrossRef]
50. Lancaster, J.E.; Shaw, M.L. Metabolism of γ-glutamyl peptides during development, storage and sprouting of onion bulbs. *Phytochemistry* **1991**, *30*, 2857–2859. [CrossRef]
51. Hughes, J.; Tregova, A.; Tomsett, A.B.; Jones, M.G.; Cosstick, R.; Collin, H.A. Synthesis of the flavour precursor, alliin, in garlic tissue cultures. *Phytochemistry* **2005**, *66*, 187–194. [CrossRef] [PubMed]
52. Mitrova, K.; Svoboda, P.; Milella, L.; Ovesna, J. Alliinase and cysteine synthase transcription in developing garlic (*Allium sativum* L.) over time. *Food Chem.* **2018**, *251*, 103–109. [CrossRef] [PubMed]

foods

MDPI

Article

Effects of $CaCl_2$ Treatment Alleviates Chilling Injury of Loquat Fruit (*Eribotrya japonica*) by Modulating ROS Homeostasis

Yuanyuan Hou, Ziying Li, Yonghua Zheng and Peng Jin *

College of Food Science and Technology, Nanjing Agricultural University, Nanjing 210095, China; 2019208027@njau.edu.cn (Y.H.); 2017108025@njau.edu.cn (Z.L.); zhengyh@njau.edu.cn (Y.Z.)
* Correspondence: pjin@njau.edu.cn; Tel.: +86-258-4395-315; Fax: +86-258-4395-618

Abstract: The effects of calcium chloride ($CaCl_2$) treatment on chilling injury (CI), reactive oxygen species (ROS) metabolism, and ascorbate-glutathione (AsA-GSH) cycle in loquat fruit at 1 °C storage for 35 d were investigated. The results indicated that $CaCl_2$ treatment remarkably suppressed the increase in browning index and firmness as well as the decrease in extractable juice rate. $CaCl_2$ treatment also decreased the production of superoxide radical ($O_2^{\bullet -}$), hydrogen peroxide (H_2O_2) content, but increased the 1,1-diphenyl-2-picrylhydrazyl (DPPH), hydroxyl radical ($OH^{\bullet}$) scavenging ability, the activities of superoxide dismutase (SOD), catalase (CAT), and their gene expressions. Moreover, compared to the control loquat fruit, $CaCl_2$-treated fruit maintained higher contents of AsA, GSH, higher levels of activities of ascorbate peroxidase (APX), glutathione reductase (GR), dehydroascorbate reductase (DHAR), and monodehydroascorbate reductase (MDHAR) and expressions of *EjAPX*, *EjGR*, *EjMDHAR*, and *EjDHAR*, but exhibited lower glutathione disulfide (GSSG) content. These results suggested that $CaCl_2$ treatment alleviated CI in loquat fruit through enhancing antioxidant enzymes activities and AsA-GSH cycle system to quench ROS.

Keywords: loquat fruit; calcium chloride; chilling injury; reactive oxygen species; ascorbate–glutathione cycle

Citation: Hou, Y.; Li, Z.; Zheng, Y.; Jin, P. Effects of $CaCl_2$ Treatment Alleviates Chilling Injury of Loquat Fruit (*Eribotrya japonica*) by Modulating ROS Homeostasis. *Foods* **2021**, *10*, 1662. https://doi.org/10.3390/foods10071662

Academic Editor: Ángel Calín-Sánchez

Received: 11 June 2021
Accepted: 15 July 2021
Published: 19 July 2021

1. Introduction

Loquat (*Eriobotrya japonica* Lindl.), as a type of non-climacteric fruit, presents a short postharvest life at ambient temperature because of its physiological deterioration and microbial decay. Refrigeration is widely adopted to retain quality and prolong the shelf life of loquat fruit. Nevertheless, loquat fruit is vulnerable to chilling injury (CI), displaying symptoms including enhanced fruit firmness, internal browning, and reduced extractable juice rate after long-term low temperature storage [1,2], which severely influences the texture and commercial value of the fruit, and eventually reduces consumer acceptance. Therefore, the exploration into the mechanism of CI in loquat fruit is of great significance.

It is well-established that the occurrence of chilling injury in higher plants is closely associated with oxidative stress resulted from excessive reactive oxygen species (ROS) such as superoxide anion ($O_2^{\bullet -}$), hydrogen peroxide (H_2O_2), and hydroxyl radical ($OH^{\bullet}$) [3]. This large quantity of ROS accumulation alter the membrane organization and cause or exacerbate lipid peroxidation, leading to damage to the cell system [4]. To protect against oxidative damage and maintain homeostasis of ROS, plants have evolved a complicated antioxidant system including enzymatic and non-enzymatic antioxidant components. Superoxide dismutase (SOD), catalase (CAT) and peroxidase (POD) are primary antioxidant enzymes, while ascorbate (AsA) and glutathione (GSH) are non-enzymatic antioxidants [5]. The ascorbate–glutathione (AsA-GSH) cycle, which mainly consists of ascorbate peroxidase (APX), glutathione reductase (GR), dehydroascorbate reductase (DHAR), and monodehydroascorbate reductase (MDHAR), plays a critical role in ROS elimination and preventing oxidative damage [6]. It has been reported that increasing the activities of antioxidant enzymes (SOD, CAT, APX, GR, MDHAR, and DHAR) to eliminate excessive ROS is beneficial for alleviating the occurrence of CI in many fruit, such as peach [7], litchi [8], and bell pepper [9]. Furthermore, previous studies have confirmed that the increased antioxidant enzymes activities (SOD, CAT, and POD) and AsA-GSH cycle and mitigated peroxidation of membrane lipids might contribute to the enhancement of chilling tolerance in loquat fruit [10,11]; nevertheless, their regulation mechanism at a molecular level remains unknown.

Calcium ion (Ca^{2+}), as an essential nutrient for fruit, plays a crucial role in building the structure of cell wall and cellular membranes [12]. Moreover, Ca^{2+} is also a crucial second messenger in plant signal transduction involving in physiological processes and the responses to various stresses [13]. Under cold stress, induced transient elevations of cytosolic Ca^{2+} are sensed by different calcium binding proteins, consequently initiating various physiological responses in the cell [14]. Dong et al. [15] reported that overexpression of the *MdCPK1a* gene increased tobacco chilling tolerance by inducing the expression of *SOD*, *CAT*, *APX* and scavenging ROS accumulation. Li et al. [16] suggested that postharvest Ca^{2+} application significantly increased Ca^{2+} and calmodulin content, and concomitantly, endogenous GABA content in 'Nanguo' pear fruit, which in turn delayed fruit browning after low-temperature storage. In addition, recent studies have reported that calcium chloride ($CaCl_2$) application can trigger antioxidant system activity and maintain ROS homeostasis to increase cold tolerance of postharvest vegetables and fruits. For instance, $CaCl_2$ application effectively alleviated peel browning caused by chilling injury of pear fruit on account of the inhibition of membrane lipid peroxidation and a higher activity and expression of SOD [17]. Wei and Zhao [18] also found that $CaCl_2$ treatment could alleviate chilling injury symptoms in winter jujube fruit by promoting the SOD, CAT, and POD activities to scavenge ROS. In green peppers, $CaCl_2$ treatment significantly suppressed ROS levels by regulating activities of SOD, POD, CAT as well as promoting the AsA-GSH cycle and, in turn, enhanced chilling tolerance [19]. For loquat fruit, a previous study demonstrated that $CaCl_2$ treatment could significantly enhance the cold tolerance of loquat fruit through regulating energy metabolism and accumulating osmotic substances [20]. However, whether the inhibitory impact of calcium application is associated with the modulation of the ROS-scavenging system for the chilling injury of loquat fruit remains unclear.

Therefore, the impact of exogenous $CaCl_2$ treatment on the ROS metabolism and AsA–GSH cycle system in loquat fruit were examined, which aimed to elucidate the antioxidant system triggered by Ca^{2+} treatment in alleviating oxidative damage during CI. The results will extend the mechanistic understanding of $CaCl_2$-inhibited chilling injury in postharvest loquat fruit, and supply a scientific basis for the application of $CaCl_2$ treatment as a useful technology to prolong the storage life of vegetables and fruits.

2. Materials and Methods

2.1. Materials and Treatments

Loquat fruit (*Eriobotrya japonica* L. cv. 'Changhong') of uniform maturity, size, and color were harvested from a commercial orchard in Fujian, China. Fruit with no mechanical damage or disease were chosen and then randomly divided into two groups (375 fruits in each group). One percent of $CaCl_2$ solution was employed to soak loquat fruit for 10 min as $CaCl_2$ treatment group, with the concentration chosen according to Li et al. [20]. Distilled water was used to immerse fruits for 10 min as control group. After treatment, all fruit were stored at 1 ± 1 °C with 90–95% relative humidity for 35 days. A total of 75 fruits from each group comprising three replicates were randomly sampled at 7 days intervals during storage. Fruit firmness was evaluated using 10 fruits, and 35 fruits were cut into small pieces, frozen using liquid nitrogen and stored at −80 °C for subsequent analysis. At each sampling day, another 30 fruits were transferred from 1 °C and kept at 20 °C for 3 days to stimulate shelf condition, and then internal browning index, firmness and extractable juice were measured. Three independent replicates were evaluated.

2.2. Measurement of Browning Index, Firmness, and Extractable Juice

The browning index was visually assayed using the reported method [20]. Browning index was scored based on a 5-grade scale (0, no visible symptoms; 1, <5% of browning; 2, 5–25% of browning; 3, 25–50% of browning; 4, >50% of browning). The result was acquired by the equation given below:

$$\text{Browning index} = \frac{\sum(\text{scale of browning}) \times (\text{number of fruit at the scale})}{(4 \times \text{total number of fruit in each tratment})} \times 10 \quad (1)$$

Fruit firmness was determined following the method of Zhang et al. [21]. Firmness from 10 loquat fruits selected randomly were assayed by a TA-XT2i texture analyzer (Stable Micro System Ltd., Surrey, UK), with a 5 mm diameter probe (test speed 1 mm s^{-1}). Firmness measurement was operated at two opposite sides of fruit, and the result was expressed as N.

Extractable juice was evaluated based on the description of Cao et al. [10]. The unit of % was used to express extractable juice rate.

2.3. Measurement of $O_2^{\bullet-}$, Generation Rate and H_2O_2 Content

The generation rate of $O_2^{\bullet-}$ and H_2O_2 content were assayed based on the description of Cao et al. [10]. The $O_2^{\bullet-}$ generation rate was acquired based on standard curve obtained from sodium nitrite and denoted as the unit of nmol g^{-1} min^{-1} on the basis of fresh weight (FW). The H_2O_2 content was acquired based on a standard curve and represented as μmol g^{-1} FW.

2.4. Measurement of 1,1-Diphenyl-2-Picrylhydrazyl (DPPH) and $OH^{\bullet}$ Radical Scavenging Capacity

Frozen tissue (2 g) was employed to determine DPPH and $OH^{\bullet}$ radical scavenging capacity according to Wang et al. [22].

For DPPH radical scavenging rate, the reaction mixture contained 0.1 mL of crude enzyme and 1.9 mL of 120 μmol L^{-1} DPPH. The result was acquired using the equation given below:

$$\text{DPPH radical scavening rate } (\%) = [(A_0 - A_1)/\ A_0] \times 100 \quad (2)$$

A_0 and A_1 indicate the absorbance of control and sample, respectively. Percentage was used for expressing DPPH radical scavenging rate.

For $OH^{\bullet}$ scavenging rate, the reaction mixture included 0.5 mL of crude enzyme, 1.5 mL of salicylic acid, 2 mL of water and 0.1 mL of 0.3% H_2O_2. Results were calculated by the following equation:

$$\text{Hydroxyl radical scavening } (\%) = [(A_0 - A_1)/\ A_0] \times 100 \quad (3)$$

A_0 and A_1 refer to absorbance of the control and samples, respectively. The unit of % was regarded as $OH^{\bullet}$ radical scavenging ability.

2.5. Determination of SOD and CAT Activities

Determination of SOD and CAT enzymes activities were carried out based on the description of Zhang et al. [21]. One unit of SOD activity was defined as the amount of enzyme causing 50% inhibition of nitroblue tetrazolium (NBT) reduction. One unit of CAT was defined as the amount of enzyme that decomposed 1 μmol of H_2O_2 min^{-1}. The unit of U g^{-1} was used to express SOD and CAT activities on the basis of fresh weight.

2.6. Measurement of Parameters Related to the AsA-GSH Cycle

Determination of AsA content was performed based on the description of Wang et al. [22], and the unit of AsA content was represented as mg g^{-1} FW based on the standard curve.

GSH and glutathione disulfide (GSSG) contents were assayed using assay kits (Solarbio, Beijing, China), according to manufacturer's instructions. GSH and GSSG contents were represented as μg g^{-1} on the basis of fresh weight.

APX activity was performed following the procedures reported by Liu et al. [8]. The APX activity was acquired from the alteration in absorbance (290 nm) for 5 min promoted by the AsA after adding H_2O_2. One unit of APX activity was defined as the amount of enzyme that led to a 0.01 variation in absorbance every minute at 290 nm, and the APX activity was represented as U g^{-1} FW.

The activity of GR was determined through monitoring the variation in absorbance at 340 nm for 2 min due to oxidation of NADPH with GSSG [8], and the results were represented as U g^{-1} FW.

Frozen tissue (2 g) was used to measure DHAR and MDHAR activities according to the description of Cao et al. [10]. DHAR activity was acquired from the variation in absorbance at 265 nm and the result was represented as U g^{-1} FW. MDHAR activity was acquired through evaluating the reduction in NADH absorbance at 340 nm and represented as U g^{-1} FW.

2.7. Real-Time Quantitative PCR (RT-qPCR) Analysis

Total RNA was extracted from loquat fruit according to an improved CTAB method [23]. First-strand cDNA was synthesized using Hifair III First Strand cDNA Synthesis Super Mix for qPCR (gDNA digester plus) (11141ES60, YEASEN, Shanghai, China). RT-qPCR analysis was conducted with a 7500 Fast Real-Time PCR System (Thermo Fisher Scientific, Waltham, MA, USA) by mixing the primers, Hieff qPCR SYBR Green Master Mix (11202ES08, YEASEN), cDNA template and RNase-free water in a total volume of 20 μL. *EjACT* was selected as the internal control to normalize the relative gene expression in each sample [24]. The relative expressions were normalized against the threshold cycle (Ct) value of the target gene and calculated with the $2^{-\Delta\Delta Ct}$ method. All analyses were conducted in triplicate, and all primer sequences were shown in Supplementary Table S1.

2.8. Statistical Analysis

Experiments were conducted at least three biological replicates and all data were expressed as mean ± standard error (SE). The statistical analyses were performed using IBM SPSS Statistics 22 (SPSS Inc., Chicago, IL, USA). The independent samples T-test was used to compare the data of control and $CaCl_2$ treatment groups. The value of $p < 0.05$ indicated significant difference.

3. Results

3.1. Effects of $CaCl_2$ Treatment on Internal Browning Index, Fruit Firmness, Extractable Juice of Loquat Fruit

Internal browning in loquat fruit occurred on the 21st day of storage and increased sharply thereafter (Figure 1A). $CaCl_2$ treatment significantly reduced the browning index under refrigeration. On day 35 of storage, browning index in $CaCl_2$-treated fruit was 47.06% lower than that in control. The firmness in loquat fruit displayed an increase trend during the entire storage period (Figure 1B). $CaCl_2$ treatment remarkably suppressed the upward trend of firmness compared to the control. Firmness in $CaCl_2$-treated fruit was 10.18% lower compared with control on the 35th day of storage. Extractable juice of the control fruit declined gradually during cold storage (Figure 1C), whereas $CaCl_2$-treated fruit displayed markedly higher extractable juice than that of the control from day 14 to day 35 of storage. These results indicated that $CaCl_2$ treatment efficiently suppressed the increase in browning index, firmness, as well as retained higher extractable juice rate, consequently inhibiting CI symptoms of loquat fruit during cold storage.

3.2. Effects of $CaCl_2$ Treatment on $O_2^{\bullet -}$ Generation Rate and H_2O_2 Content of Loquat Fruit

$O_2^{\bullet -}$ generation rate and H_2O_2 level in control loquat fruit displayed overall increments during cold storage (Figure 2). However, compared to the control, the application of $CaCl_2$ remarkably reduced the accumulation of $O_2^{\bullet -}$ and H_2O_2 throughout the whole storage time. At the end of storage, the $O_2^{\bullet -}$ generation rate and H_2O_2 content of $CaCl_2$-treated fruit were 10.00% and 14.71% lower ($p < 0.05$) than those in control fruit, respectively.

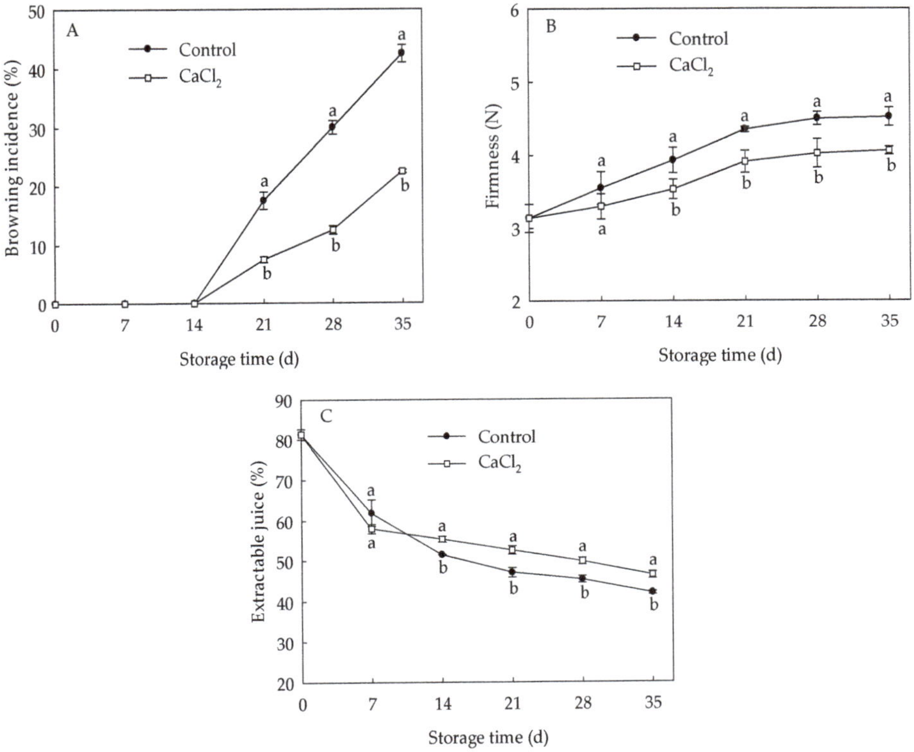

Figure 1. Browning index (**A**), firmness (**B**), and extractable juice (**C**) in loquat fruit stored at 1 °C after treatment with 1% $CaCl_2$ or water (control). Vertical bars indicate the SE of mean ($n = 3$). The symbols a and b represent a significant difference in the two treatments ($p < 0.05$).

3.3. Effects of $CaCl_2$ Treatment on DPPH and $OH^{\bullet}$ Radical Scavenging Capacity of Loquat Fruit

DPPH radical scavenging rate of control loquat fruit declined at the first 7 days of storage, and increased gradually from day 7 to day 21, then maintained relatively stable, while the $CaCl_2$-treated fruit exhibited a gradually upward trend (Figure 3A). $CaCl_2$ treatment maintained higher DPPH radical scavenging capacity within the entire storage period, and the DPPH radical scavenging rate was 27.85% higher in $CaCl_2$-treated fruit than that of the control on the 35th day. The $OH^{\bullet}$ radical scavenging rate in control loquat fruit increased before day 7 of storage, but declined during day 7 to day 28, then displayed a slight increase (Figure 3B). However, the $OH^{\bullet}$ radical scavenging rate in the fruit treated with $CaCl_2$ was higher compared to control fruit within the whole storage time, and presented a markedly ($p < 0.05$) higher $OH^{\bullet}$ radical scavenging rate during 14 d to 28 d of storage.

3.4. Effects of $CaCl_2$ Treatment on CAT, SOD Activities and EjCAT, EjSOD Expressions of Loquat Fruit

The SOD activity of loquat fruit rose quickly from 0 to 14 d, but gradually decreased thereafter (Figure 4A). The CAT activity in loquat fruit dropped slowly from day 0 to day 7 of storage, but quickly rose from 7 d to 14 d, followed by a gradual declination (Figure 4C). $CaCl_2$ treatment enhanced SOD and CAT activities throughout the whole storage period, with 21.61% and 21.84% higher than those in control on day 21, respectively. Expression levels of *EjSOD* and *EjCAT* presented a similar trend with that of SOD and CAT enzymes activities (Figure 4B,D). $CaCl_2$ treatment up regulated the expression of *EjSOD* and *EjCAT*, which were approximate 2.80-fold and 17.83% higher on the 21st day of storage compared to control, respectively.

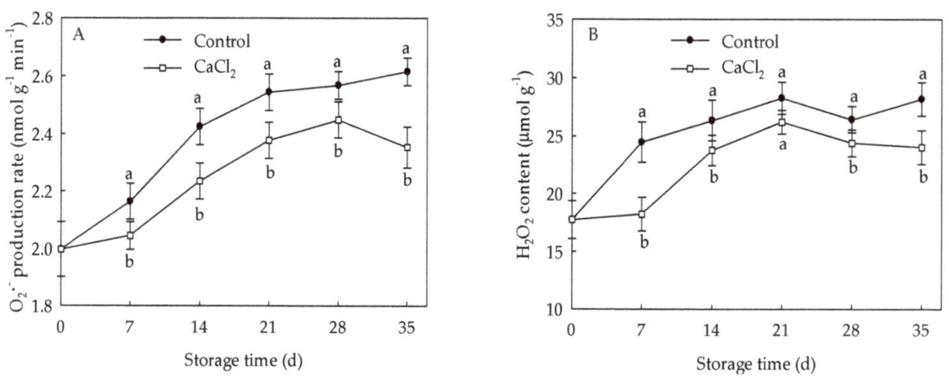

Figure 2. $O_2^{\bullet-}$ generation rate (**A**) and H_2O_2 content (**B**) in loquat fruit stored at 1 °C after treatment with 1% $CaCl_2$ or water (control). Vertical bars indicate the SE of mean ($n = 3$). The symbols a and b represent a significant difference in the two treatments ($p < 0.05$).

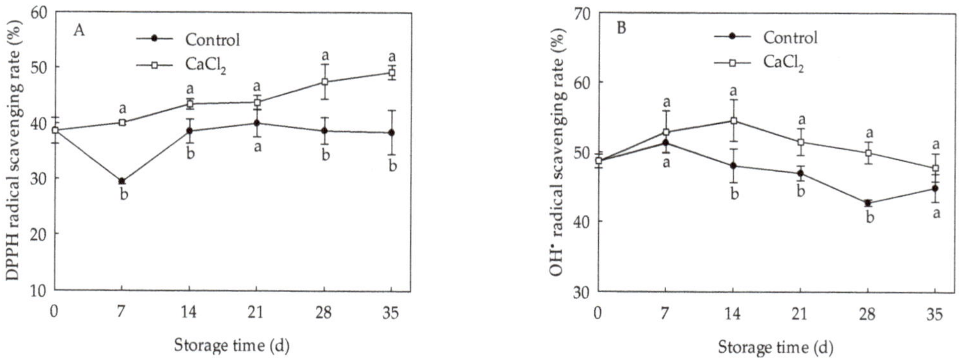

Figure 3. DPPH (**A**) and $OH^{\bullet}$ (**B**) radical scavenging capacity in loquat fruit stored at 1 °C after treatment with 1% $CaCl_2$ or water (control). Vertical bars indicate the SE of mean ($n = 3$). The symbols a and b represent a significant difference in the two treatments ($p < 0.05$).

3.5. Effects of $CaCl_2$ Treatment on Contents of AsA, GSH, GSSG and GSH/GSSG Ratio of Loquat Fruit

The content of AsA in loquat fruit continuously declined with the extension of storage time (Figure 5A), whereas GSH content rose quickly from day 0 to day 21 of storage, then dropped slowly during 21–28 d, followed by a sharp increase (Figure 5B). Change in content of GSSG in loquat fruit was similar to that of AsA content (Figure 5C). On the contrary, GSH/GSSG in loquat fruit displayed a steady increase trend during cold storage

(Figure 5D). $CaCl_2$ treatment maintained higher AsA content and GSH content compared to the control, whereas GSSG content was lower in $CaCl_2$-treated fruit than that in the control group. $CaCl_2$ treatment remarkably increased GSH/GSSG within the entire storage time, and the GSH/GSSG in $CaCl_2$-treated fruit was 34.92% higher on the 35th day of storage compared with control.

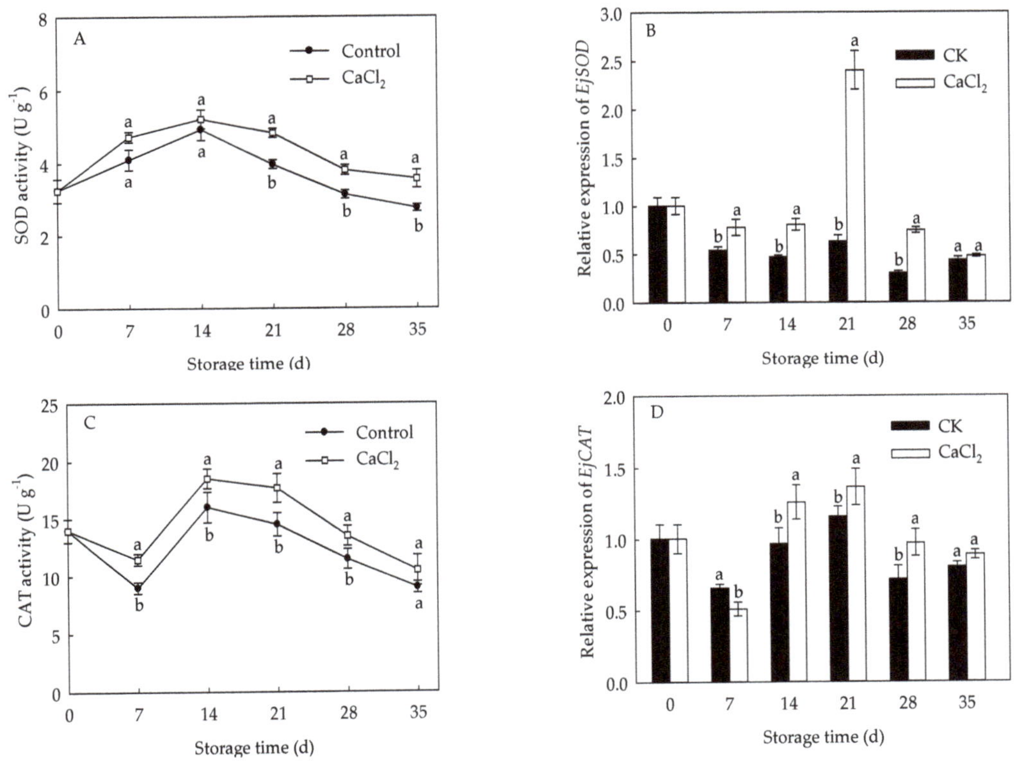

Figure 4. SOD activity (**A**), *EjSOD* expression (**B**), CAT activity (**C**) and *EjCAT* expression (**D**) in loquat fruit stored at 1 °C after treatment with 1% $CaCl_2$ or water (control). Vertical bars indicate the SE of mean ($n = 3$). The symbols a and b represent a significant difference in the two treatments ($p < 0.05$).

3.6. Effects of $CaCl_2$ Treatment on Enzyme Activity and Gene Expression of APX, GR, DHAR, and MDHAR of Loquat Fruit

The APX activity in loquat fruit reached the peak on the 21st day of storage, and dropped quickly afterwards (Figure 6A). $CaCl_2$-treated fruit exhibited higher APX activity compared to control fruit. Change of GR activity in loquat fruit displayed a similar trend with APX activity, but the maximum value reached on the 28th day (Figure 6C). Compared to control group, $CaCl_2$ treatment maintained higher GR activity during cold storage, with a remarkable discrepancy from day 14 to day 35, except day 21 of storage.

The DHAR activity in control loquat fruit rose at first 14 d of storage, then declined during 14–28 d, followed by a slight increase (Figure 6E). However, $CaCl_2$-treated fruit exhibited remarkably higher DHAR activity from day 14 to day 35 compared to the control fruit. MDHAR activity of control fruit decreased sharply initially, and rose during day 7 to day 21, then a slight drop during day 21 to day 35 (Figure 6G). $CaCl_2$ treatment displayed remarkably higher MDHAR activity than that in control group during day 7 to day 35, except day 28.

The expression of *EjAPX* and *EjDHAR* in control loquat fruit generally showed a downward trend, but the $CaCl_2$ treatment remained higher expression of *EjAPX* and *EjDHAR* throughout the entire storage period (Figure 6B,F). On day 21 of storage, *EjDHAR* expression in $CaCl_2$-treated loquat fruit was about 6.82-fold of control. Expression of *EjGR* and *EjMDHAR* in loquat fruit increased and peaked at 21 d of storage, followed by an overall decline (Figure 6D,H). Compared with control fruit, $CaCl_2$ treatment up regulated the expression of *EjGR* and *EjMDHAR* within the whole storage time. Expression of *EjGR* and *EjMDHAR* in the fruit treated with $CaCl_2$ were 3.08-fold and 3.55-fold of those in control on the 21st day of storage, respectively.

4. Discussion

Chilling injury, manifested as internal browning, an unusual increase in firmness, and juiceless pulp, is a primary problem that significantly impacts the quality of loquat fruit during cold storage [10]. It is reported that Ca^{2+} acts as an important second messenger fulfilling a crucial role in protection against chilling injury in plants [13]. Previous studies have confirmed that exogenous $CaCl_2$ treatment improved chilling tolerance in loquat fruit under low temperature storage [20]. In the present study, $CaCl_2$ treatment efficiently suppressed the increase in browning index and firmness, and maintained higher extractable juice in loquat fruit (Figure 1), demonstrating $CaCl_2$ exerts a positive influence in enhancing chilling tolerance in harvested loquat fruit. Similar result was also obtained in pear fruit [17] and pineapple fruit [25].

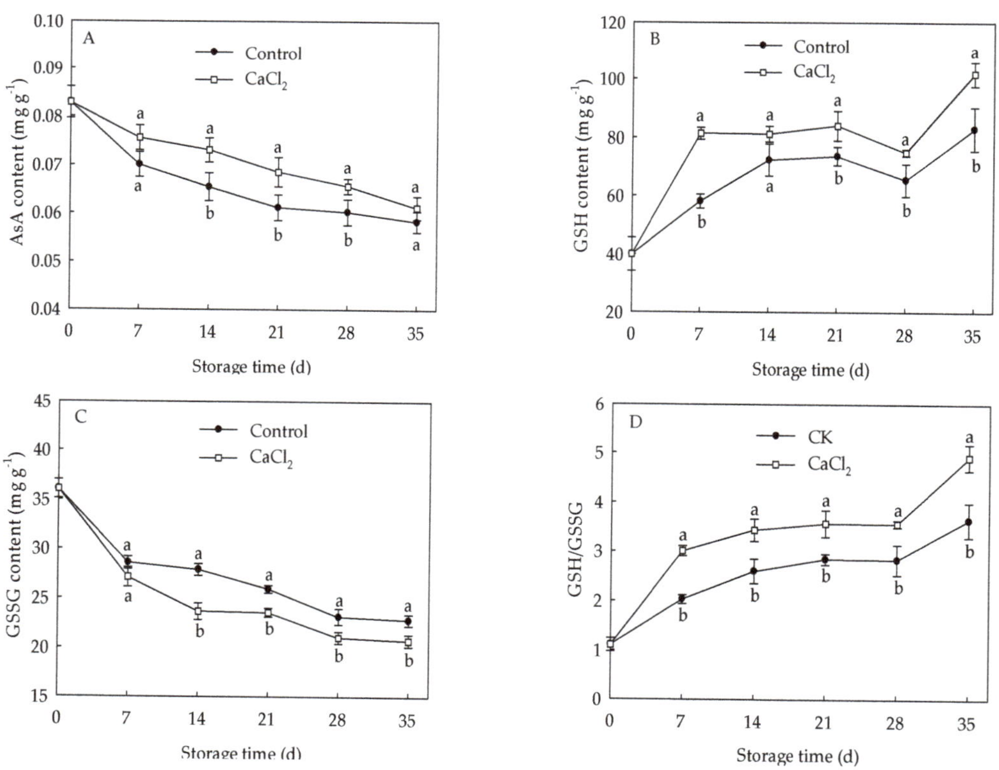

Figure 5. AsA (**A**), GSH (**B**), GSSG (**C**) contents and GSH/GSSG (**D**) in loquat fruit stored at 1 °C after treatment with 1% $CaCl_2$ or water (control). Vertical bars indicate the SE of mean (n = 3). The symbols a and b represent a significant difference in the two treatments ($p < 0.05$).

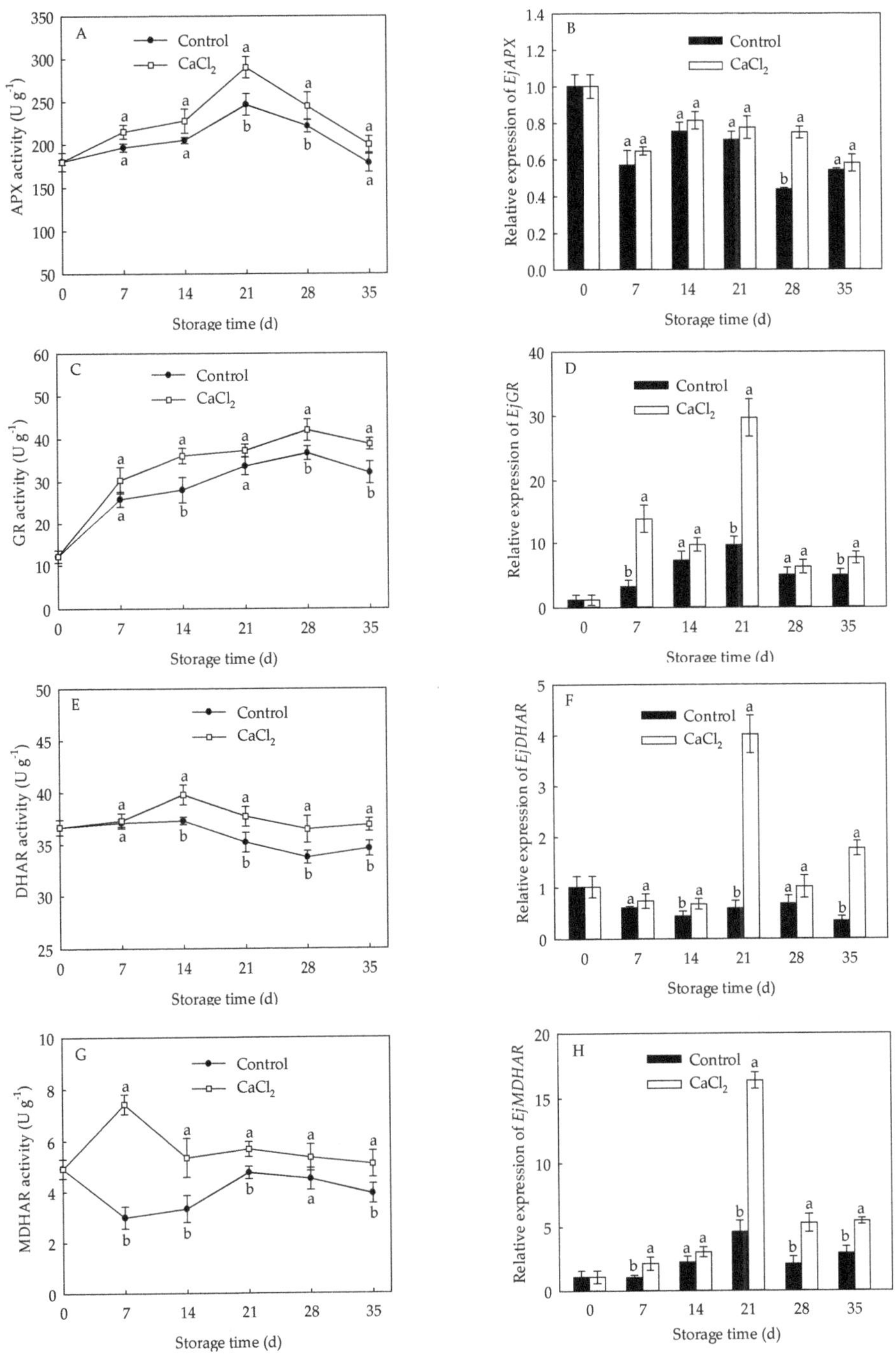

Figure 6. APX activity (**A**), *EjAPX* expression (**B**), GR activity (**C**), *EjGR* expression (**D**), DHAR activity (**E**), *EjDHAR* expression (**F**), MDHAR activity (**G**), and *EjMDHAR* expression (**H**) in loquat fruit stored at 1 °C after treatment with 1% $CaCl_2$ or water (control). Vertical bars indicate the SE of mean ($n = 3$). The symbols a and b represent a significant difference in the two treatments ($p < 0.05$).

Destabilization of the cell membrane has been well-known as the primary reason of chilling injury in plants. The overproduction of ROS, including $O_2^{\bullet-}$, H_2O_2 and hydroxyl radical, may result in the occurrence of CI [26,27]. SOD, CAT, and APX are considered to be vital ROS-eliminating enzymes which involve an alleviation of chilling injury [9]. SOD can convert the overproduced $O_2^{\bullet-}$ into H_2O_2, while CAT and APX catalyze the H_2O_2 into H_2O and O_2 [28]. Antioxidant enzymes with higher activity and their integrated action have been demonstrated to be a part of the mechanism involved in the inhibition of oxidative damage and enhancement of cold tolerance in pear [29], blood orange [26], and banana [30]. In the current study, the accumulation of $O_2^{\bullet-}$ and H_2O_2 in loquat fruit (Figure 2) demonstrated the membrane damage from accumulation of ROS under chilling stress, and $CaCl_2$ treatment efficiently reduced ROS accumulation, leading to a lower membrane peroxidation in loquat fruit [20]. Compared with the control fruit, $CaCl_2$-treated loquat fruit maintained higher DPPH and $OH^{\bullet}$ radical scavenging ability (Figure 3), which are two crucial parameters on the antioxidant capacity of plants [31]. Meanwhile, the SOD and CAT activities in loquat fruit rose during the early stage of storage, and $CaCl_2$ treatment enhanced SOD and CAT activities within the whole storage period (Figure 4A,C). Moreover, the expression of *EjSOD* and *EjCAT* was also up regulated by $CaCl_2$ treatment (Figure 4B,D). The increased activities and gene expressions of SOD and CAT might explain the lower levels of $O_2^{\bullet-}$ and H_2O_2 in $CaCl_2$-treated fruit. Zhang et al. [17] also reported that postharvest $CaCl_2$ treatment promoted the activity and expression of SOD to against oxidative damage, inhibiting the browning in pear fruit after cold storage. Shi et al. [32] reported that $CaCl_2$ treatment increased SOD and CAT activities to modulate redox homeostasis, contributing to improvement of chilling tolerance in bermudagrass. Similar results have also been confirmed in loquat fruit treated by methyl jasmonate [27]. These results indicated that the effect of $CaCl_2$ on modulating the ROS homeostasis of loquat fruit was correlated to enhanced enzyme activities as well as gene expressions of SOD and CAT which, in turn, alleviated the occurrence of chilling injury.

The ASA–GSH cycle acts as a crucial antioxidant system to substantially scavenges H_2O against oxidative damage, and APX, GR, DHAR, and MDHAR are key enzymes in this cycle [9]. AsA and GSH are products of the ASA-GSH cycle and important non-enzymatic antioxidants that are capable of directly or indirectly quenching ROS [7]. AsA is oxidized to the monodehydroascorbate (MDHA) radical, while APX converts H_2O_2 into the H_2O with the help of AsA as an electron donor. MDHA regenerates AsA by MDHAR and is spontaneously converted to dehydroascorbate (DHA). DHA is reduced to AsA again through GSH, which leads to its oxidation to produce GSSG by DHAR. GSH regenerate from GSSG with catalyzation of GR [33]. Previous studies have demonstrated that enhanced AsA, GSH contents, and GSH/GSSG ratio are beneficial for the resistance to oxidative damage and improvement of chilling tolerance in plants [34,35]. Zhang et al. [19] confirmed that Ca^{2+} treatment could increase the activities of APX and GR to activate AsA-GSH cycle, and to effectively inhibit the increase in H_2O_2, thus, maintaining redox balance to better alleviate the occurrence of oxidative damage to green peppers under cold stress. In the current study, $CaCl_2$-treated loquat fruit displayed higher levels of AsA and GSH contents (Figure 5A,B), lower GSSG content (Figure 5C), and retained higher GSH/GSSG ratio (Figure 5D) compared with the control fruit. Furthermore, $CaCl_2$ enhanced the APX, GR, DHAR, and MDHAR activities in loquat fruit under cold storage (Figure 6A,C,E,G), which was conducive to maintaining a higher content of AsA, GSH and GSH/GSSG ratio. These findings were consistent with the result that ROS levels in $CaCl_2$-treated fruit was lower than that in control (Figure 2). Similar results that AsA-GSH cycle played a crucial role in maintaining redox balance by loquat fruit under chilling injury were also reported by Cao et al. [10].

Additionally, the numerous reports showed that the increased expression of *APX*, *GR*, *DHAR*, *MDHAR* could alleviate oxidative injury due to their role in modulation of AsA–GSH cycle system in plants [33]. Song et al. [7] demonstrated that the mitigation of CI in peach fruit treated by hypobaric treatment was associated with its influence on activating the AsA-GSH cycle system through inducing *GR*, *MDHAR1*, and *APX* gene expressions to avoid oxidative damage. In the current study, $CaCl_2$ treatment up regulated the *EjAPX*, *EjGR*, *EjDHAR*, and *EjMDHR* expressions in loquat fruit (Figure 6B,D,F,H), which exerted positive influence in the maintenance of high activities of antioxidant enzymes involved in AsA–GSH cycle, resulting in the inhibition of ROS accumulation and chilling injury symptoms. It was in agreement with previous work which stated that the up-regulated *CaAPX1*, *CaMDHAR1*, and *CaDHAR1* expressions in bell peppers by GSH treatment promoted activities of AsA-GSH cycle enzymes and regenerated AsA and GSH, consequently bringing about lower oxidative damage and alleviation of chilling injury [9]. Thus, we conclude that $CaCl_2$ treatment could mitigate oxidative damage by activating AsA-GSH cycle system in loquat fruit during chilling injury.

5. Conclusions

In conclusion, $CaCl_2$ treatment effectively enhanced cold tolerance and alleviated the occurrence of CI in loquat fruit. The inhibition in occurrence of CI by $CaCl_2$ treatment might be ascribed to its capacity to reduce ROS accumulation by triggering of the antioxidant system including enzymatic and non-enzymatic antioxidant activities, thereby reducing cell membrane oxidative damage. These findings provide insight into the improvement of cold tolerance in loquat fruit through modulating ROS homeostasis by $CaCl_2$ treatment. However, further molecular evidence is needed to support an explanation of how Ca^{2+} triggers low-temperature tolerance.

Supplementary Materials: The following are available online at https://www.mdpi.com/article/10.3390/foods10071662/s1, Table S1. Primer sequences for Real-time PCR analysis.

Author Contributions: Conceptualization, Y.H. and P.J.; data curation, Y.H. and Z.L.; funding acquisition, P.J.; investigation, Y.H. and Z.L.; methodology, Y.H. and Z.L.; supervision, P.J.; validation, P.J. and Y.Z.; writing—original draft, Y.H. All authors have read and agreed to the published version of the manuscript.

Funding: This study was supported by the National Natural Science Foundation of China (No.31972125, 31671901).

Institutional Review Board Statement: Not applicable.

Informed Consent Statement: Not applicable.

Data Availability Statement: Not applicable.

Acknowledgments: The authors gratefully acknowledge laboratory colleagues for their help.

Conflicts of Interest: The authors declare no conflict of interest.

References

1. Zheng, Y.H.; Li, S.Y.; Xi, Y.F. Changes of cell wall substances in relation to flesh woodiness in cold-stored loquat fruits. *Acta Phytophysiol. Sin.* **2000**, *26*, 306–310. [CrossRef]
2. Cai, C.; Xu, C.J.; Li, X.; Ferguson, I.B.; Chen, K.S. Accumulation of lignin in relation to change in activities of lignification enzymes in loquat fruit flesh after harvest. *Postharvest Boil. Technol.* **2006**, *40*, 163–169. [CrossRef]
3. Dreyer, A.; Dietz, K.J. Reactive Oxygen species and the redox-regulatory network in cold stress acclimation. *Antioxidants* **2018**, *7*, 169. [CrossRef]
4. You, J.; Chan, Z.L. ROS regulation during abiotic stress responses in crop plants. *Front. Plant Sci.* **2015**, *6*, 1092. [CrossRef]
5. Gill, S.S.; Tuteja, N. Reactive oxygen species and antioxidant machinery in abiotic stress tolerance in crop plants. *Plant Physiol. Biochem.* **2010**, *48*, 909–930. [CrossRef] [PubMed]
6. Zhao, H.; Ye, L.; Wang, Y.; Zhou, X.; Yang, J.; Wang, J.; Cao, K.; Zou, Z. Melatonin increases the chilling tolerance of chloroplast in cucumber seedlings by regulating photosynthetic electron flux and the ascorbate-glutathione cycle. *Front. Plant Sci.* **2016**, *7*, 1814. [CrossRef]
7. Song, L.L.; Wang, J.H.; Shafi, M.; Liu, Y.; Wang, J.; Wu, J.S.; Wu, A.M. Hypobaric treatment effects on chilling injury, mitochondrial dysfunction, and the ascorbate–glutathione (AsA-GSH) cycle in postharvest peach fruit. *J. Agric. Food Chem.* **2016**, *64*, 4665–4674. [CrossRef]

8. Liu, J.L.; Sun, J.H.; Pan, Y.G.; Yun, Z.; Zhang, Z.K.; Guoxiang Jiang, G.X.; Jiang, Y.M. Endogenous melatonin generation plays a positive role in chilling tolerance in relation to redox homeostasis in litchi fruit during refrigeration. *Postharvest Boil. Technol.* **2021**, *178*, 111554. [CrossRef]
9. Yao, M.; Ge, W.; Zhou, Q.; Zhou, X.; Luo, M.; Zhao, Y.; Wei, B.; Ji, S. Exogenous glutathione alleviates chilling injury in postharvest bell pepper by modulating the ascorbate-glutathione (AsA-GSH) cycle. *Food Chem.* **2021**, *352*, 129458. [CrossRef] [PubMed]
10. Cao, S.; Yang, Z.; Cai, Y.; Zheng, Y. Fatty acid composition and antioxidant system in relation to susceptibility of loquat fruit to chilling injury. *Food Chem.* **2011**, *127*, 1777–1783. [CrossRef]
11. Cai, Y.; Cao, S.; Yang, Z.; Zheng, Y. MeJA regulates enzymes involved in ascorbic acid and glutathione metabolism and improves chilling tolerance in loquat fruit. *Postharvest Boil. Technol.* **2011**, *59*, 324–326. [CrossRef]
12. Hocking, B.; Tyerman, S.D.; Burton, R.A.; Gilliham, M. Fruit calcium: Transport and physiology. *Front. Plant Sci.* **2016**, *7*, 569. [CrossRef]
13. Ranty, B.; Aldon, D.; Cotelle, V.; Galaud, J.P.; Thuleau, P.; Mazars, C. Calcium sensors as key hubs in plant responses to biotic and abiotic stresses. *Front. Plant Sci.* **2016**, *7*, 327. [CrossRef] [PubMed]
14. Yuan, P.; Yang, T.; Poovaiah, B.W. Calcium signaling-mediated plant response to cold stress. *Int. J. Mol. Sci.* **2018**, *19*, 3896. [CrossRef] [PubMed]
15. Dong, H.; Wu, C.; Luo, C.; Wei, M.; Qu, S.; Wang, S. Overexpression of *MdCPK1a* gene, a calcium dependent protein kinase in apple, increase tobacco cold tolerance via scavenging ROS accumulation. *PLoS ONE* **2020**, *15*, e0242139. [CrossRef] [PubMed]
16. Li, J.; Zhou, Q.; Zhou, X.; Wei, B.; Ji, S. Calcium treatment alleviates pericarp browning of 'nanguo' pears by regulating the GABA shunt after cold storage. *Front. Plant Sci.* **2020**, *11*, 580986. [CrossRef] [PubMed]
17. Zhang, L.; Wang, J.; Zhou, B.; Li, G.; Liu, Y.; Xia, X.L.; Xiao, Z.G.; Fei, L.; Ji, S.J. Calcium inhibited peel browning by regulating enzymes in membrane metabolism of 'Nanguo' pears during post-ripeness after refrigerated storage. *Sci. Hortic.* **2019**, *244*, 15–21. [CrossRef]
18. Wei, D.; Zhao, X. Calcium maintained higher quality and enhanced resistance against chilling stress by regulating enzymes in reactive oxygen and biofilm metabolism of chinese winter jujube fruit. *J. Food Biochem.* **2020**, *44*, e13161. [CrossRef] [PubMed]
19. Zhang, X.Y.; Ma, M.J.; Ye, B.; Liu, L.; Ji, S.J. Calcium ion improves cold resistance of green peppers (*Capsicum annuum* L.) by regulating the activity of protective enzymes and membrane lipid composition. *Sci. Hortic.* **2021**, *277*, 109789. [CrossRef]
20. Li, Z.Y.; Wang, L.; Xie, B.; Hu, S.Q.; Zheng, Y.H.; Jin, P. Effects of exogenous calcium and calcium chelant on cold tolerance of postharvest loquat fruit. *Sci. Hortic.* **2020**, *269*, 109391. [CrossRef]
21. Zhang, Y.; Jin, P.; Huang, Y.P.; Shan, T.M.; Wang, L.; Li, Y.Y.; Zheng, Y.H. Effect of hot water combined with glycine betaine alleviates chilling injury in cold-stored loquat fruit. *Postharvest Boil. Technol.* **2016**, *118*, 141–147. [CrossRef]
22. Wang, L.; Shao, S.; Madebo, M.P.; Hou, Y.; Zheng, Y.; Jin, P. Effect of nano-SiO_2 packing on postharvest quality and antioxidant capacity of loquat fruit under ambient temperature storage. *Food Chem.* **2020**, *315*, 126295. [CrossRef] [PubMed]
23. Asif, M.H.; Dhawan, P.; Math, P. A simple procedure for the isolation of high quality RNA from ripening banana fruit. *Plant Mol. Biol. Rep.* **2000**, *18*, 109–115. [CrossRef]
24. Zeng, J.K.; Li, X.; Zhang, J.; Ge, H.; Yin, X.R.; Chen, K.S. Regulation of loquat fruit low temperature response and lignification involves interaction of heat shock factors and genes associated with lignin biosynthesis. *Plant Cell Environ.* **2016**, *39*, 1780–1789. [CrossRef]
25. Youryon, P.; Supapvanich, S.; Kongtrakool, P.; Wongs-Aree, C. Calcium chloride and calcium gluconate peduncle infiltrations alleviate the internal browning of queen pineapple in refrigerated storage. *Hortic. Environ. Biotechnol.* **2018**, *59*, 205–213. [CrossRef]
26. Habibi, F.; Ramezanian, A.; Guillén, F.; Martínez-Romero, D.; Serrano, M.; Valero, D. Susceptibility of blood orange cultivars to chilling injury based on antioxidant system, physiological and biochemical responses at different storage temperatures. *Foods* **2020**, *9*, 1609. [CrossRef] [PubMed]
27. Cao, S.; Zheng, Y.; Wang, K.; Peng, J.; Rui, H. Methyl jasmonate reduces chilling injury and enhances antioxidant enzyme activity in postharvest loquat fruit. *Food Chem.* **2009**, *115*, 1458–1463. [CrossRef]
28. Mittler, R. Oxidative stress, antioxidants and stress tolerance. *Trends Plant Sci.* **2002**, *7*, 405–410. [CrossRef]
29. Sun, H.; Luo, M.L.; Zhou, X.; Zhou, Q.; Ji, S.J. Exogenous glycine betaine treatment alleviates low temperature-induced pericarp browning of 'nanguo' pears by regulating antioxidant enzymes and proline metabolism. *Food Chem.* **2019**, *306*, 125626. [CrossRef]
30. Hao, J.; Li, X.; Xu, G.; Huo, Y.; Yang, H. Exogenous progesterone treatment alleviates chilling injury in postharvest banana fruit associated with induction of alternative oxidase and antioxidant defense. *Food Chem.* **2019**, *286*, 329–337. [CrossRef]
31. Tan, X.L.; Zhao, Y.T.; Shan, W.; Kuang, J.F.; Chen, J.Y. Melatonin delays leaf senescence of postharvest chinese flowering cabbage through ros homeostasis. *Food Res. Int.* **2020**, *138*, 109790. [CrossRef] [PubMed]
32. Shi, H.; Ye, T.; Ba, O.Z.; Liu, X.; Chan, Z. Comparative proteomic and metabolomic analyses reveal mechanisms of improved cold stress tolerance in bermudagrass (*cynodon dactylon* (L.) Pers.) by exogenous calcium. *J. Integr. Plant Biol.* **2014**, *56*, 1064–1079. [CrossRef] [PubMed]
33. Hasanuzzaman, M.; Bhuyan, M.H.M.B.; Anee, T.I.; Parvin, K.; Nahar, K.; Mahmud, J.A.; Fujita, M. Regulation of ascorbate-glutathione pathway in mitigating oxidative damage in plants under abiotic stress. *Antioxidants* **2019**, *8*, 384. [CrossRef]
34. Chen, Z.Y.; Wang, Y.T.; Pan, X.B.; Xi, Z.M. Amelioration of cold-induced oxidative stress by exogenous 24-epibrassinolide treatment in grapevine seedlings: Toward regulating the ascorbate–glutathione cycle. *Sci. Hortic.* **2019**, *244*, 379–387. [CrossRef]
35. Huan, C.; Han, S.A.; Jiang, L.; An, X.J.; Yu, M.L.; Xu, Y.; Ma, R.J.; Yu, Z.F. Postharvest hot air and hot water treatments affect the antioxidant system in peach fruit during refrigerated storage. *Postharvest Boil. Technol.* **2017**, *126*, 1–14. [CrossRef]

foods

MDPI

Article

Melatonin Treatment Improves Postharvest Preservation and Resistance of Guava Fruit (*Psidium guajava* L.)

Silin Fan [1], Tiantian Xiong [2], Qiumei Lei [1], Qinqin Tan [1], Jiahui Cai [1], Zunyang Song [1], Meiyan Yang [1], Weixin Chen [1], Xueping Li [1] and Xiaoyang Zhu [1,*]

[1] Guangdong Provincial Key Laboratory of Postharvest Science of Fruits and Vegetables/Engineering Research Center for Postharvest Technology of Horticultural Crops in South China, Ministry of Education, College of Horticulture, South China Agricultural University, Guangzhou 510642, China; 1575644390@foxmail.com (S.F.); 1198960312@foxmail.com (Q.L.); 1621251540@foxmail.com (Q.T.); jh.cai_chn@outlook.com (J.C.); songzunyang@163.com (Z.S.); ymy@scau.edu.cn (M.Y.); wxchen@scau.edu.cn (W.C.); lxp88@scau.edu.cn (X.L.)

[2] Key Laboratory of Ecology and Environmental Science in Guangdong Higher Education, School of Life Science, South China Normal University, Guangzhou 510631, China; xiongtiantian@m.scnu.edu.cn

* Correspondence: xiaoyang_zhu@scau.edu.cn; Tel.: +86-20-85288280; Fax: +86-20-85288280

Abstract: Guava fruit has a short postharvest shelf life at room temperature. Melatonin is widely used for preservation of various postharvest fruit and vegetables. In this study, an optimal melatonin treatment (600 $\mu mol \cdot L^{-1}$, 2 h) was identified, which effectively delayed fruit softening and reduced the incidence of anthracnose on guava fruit. Melatonin effectively enhanced the antioxidant capacity and reduced the oxidative damage to the fruit by reducing the contents of superoxide anions, hydrogen peroxide and malondialdehyde; improving the overall antioxidant capacity and enhancing the enzymatic antioxidants and non-enzymatic antioxidants. Melatonin significantly enhanced the activities of catalase, superoxide dismutase, ascorbate peroxidase and glutathione reductase. The contents of total flavonoids and ascorbic acid were maintained by melatonin. This treatment also enhanced the defense-related enzymatic activities of chitinase and phenylpropanoid pathway enzymes, including phenylalanine ammonia lyase and 4-coumaric acid-CoA-ligase. The activities of lipase, lipoxygenase and phospholipase D related to lipid metabolism were repressed by melatonin. These results showed that exogenous melatonin can maintain the quality of guava fruit and enhance its resistance to disease by improving the antioxidant and defense systems of the fruit.

Keywords: ripening; reactive oxygen species; enzymatic activities; anthracnose; shelf-life

Citation: Fan, S.; Xiong, T.; Lei, Q.; Tan, Q.; Cai, J.; Song, Z.; Yang, M.; Chen, W.; Li, X.; Zhu, X. Melatonin Treatment Improves Postharvest Preservation and Resistance of Guava Fruit (*Psidium guajava* L.). *Foods* **2022**, *11*, 262. https://doi.org/10.3390/foods11030262

Academic Editor: Peng Jin

Received: 14 December 2021
Accepted: 15 January 2022
Published: 19 January 2022

Publisher's Note: MDPI stays neutral with regard to jurisdictional claims in published maps and institutional affiliations.

1. Introduction

Guava (*Psidium guajava* L.) is native to tropical America from Mexico to Peru, and now a commercially important fruit and medicinal plant in many tropical and subtropical countries [1]. Countries producing and consuming most guava include India, Mexico, Pakistan, Brazil, Egypt, Thailand, Colombia [1], as well as China in recent years. The significance of guava fruit may be owing to its high production, remarkable nutritional properties and enchanting taste, since it contains five times as much ascorbic acid as citrus fruit and also contains various essential bioactive compounds [2]. Guava has high levels of dietary fiber, pectin, antioxidants, vitamins and mineral content compared with other fruit [2,3]. However, guava fruit with thin peel and soft texture are highly susceptible to chilling stress, mechanical injury and pathogens. Therefore, it has a short postharvest life, which severely limits its consumption and distribution to domestic markets [4]. The fruit soften rapidly, so they tend to bruise and over-ripen, which causes severe losses during the postharvest period. To date, various postharvest handling techniques have been utilized to regulate the ripening of fruit and reduce losses from postharvest physiological disorders and diseases. Low temperature storage is a well-known and common method to extend the storage period of guava fruit. However, storage under 10 °C can cause serious chilling injury (CI) to guava fruit, including the browning or discoloration of the peel and pulp, abnormal ripening and an increase in the incidence of anthracnose caused by *Colletotrichum gloeosporioides* [5]. Anthracnose is one of the principal postharvest diseases of several tropical and subtropical fruit, including papayas, mangoes, bananas, apples, passion fruit, guavas, grapes and citrus [6]. As the most common fungal disease of guava fruit (*Lim and Manicom*), anthracnose causes severe postharvest losses of these fruit [1]. The anthracnose disease commonly infects quiescently, initiating infection during flowering and remaining latent until it is activated by ripening during fruit storage, transport and shelf life [6,7]. Different technologies have been explored to reduce the incidence of anthracnose in guava fruit. 1-methylcyclopropene (1-MCP), an effective inhibitor of ethylene, is widely used to extend the shelf life of fruit. 1-MCP has a pronounced effect on fruit ripening and significantly maintains high fruit firmness, maintains fruit quality and reduces the incidence of decay [4]. Other commonly used techniques, such as edible coating treatments, modified atmosphere, and controlled atmosphere storage, ascorbic acid treatment, gamma irradiation, fungicides, calcium chloride and nitric oxide, can reduce the fruit peel browning index, maintain the fruit quality and extend the shelf life of guava fruit [1,3,8–13].

Melatonin (N-acetyl-5-methoxytryptamine) is a multifunctional signaling molecule that is ubiquitously distributed in various plant species [14]. Melatonin is involved in various physiological activities in plants, including seed and root development [15], plant growth and flowering [16], fruit development [17], fruit ripening [18], senescence [19] and biotic and abiotic stress responses [20,21]. As a safe and nontoxic substance, melatonin is widely studied in postharvest preservation for its role in improving the storage life and quality of fruit and vegetables, as well as enhancing crop production and ensuring food safety in an environmentally friendly manner. Various studies have proven that the application of exogenous melatonin significantly delayed the postharvest ripening of some fruit, such as apples [22], sweet cherries [23], bananas [19,24], pears [25], peaches [26] and mangoes [27]. Exogenous treatment with melatonin can also enhance the resistance to disease and reduce the incidence of decay of various fruit during postharvest storage, including litchis [28], strawberries [29], kiwifruits [30] and peaches [26].

This study aims to explore the effect of melatonin on guava fruit ripening and quality during the postharvest storage periods. The results showed that treatment with melatonin significantly delayed fruit ripening and repressed the incidence of anthracnose during the postharvest periods. In particular, treatment with melatonin effectively enhanced the antioxidant capacity and reduced the oxidative damage to the fruit. These results provide technical and theoretical guidance for the practical application of melatonin in the postharvest stage of guava fruit during storage and marketing.

2. Materials and Methods

2.1. Plant Materials and Treatments

Guava fruit (*Psidium guajava* L. 'Zhenzhu') at the commercial stage (mature-green) were harvested from a local commercial plantation of Baoshen Farm, Nansha district, Guangzhou, South China, and immediately transported to the laboratory. Fruit without blemishes that were uniformly sized were selected, cleaned and dipped in a 0.2% solution of sodium hypochlorite solution for 10 min. After air-drying under the room temperature (22 ± 2 °C) for 2 h (h), selected fruit were soaked in different concentrations of melatonin (Shanghai Shenggong Biology Limited Company, Shanghai, China) for 2 h. The melatonin was dissolved in ethanol and then diluted to 0 $\mu mol \cdot L^{-1}$, 100 $\mu mol \cdot L^{-1}$, 400 $\mu mol \cdot L^{-1}$, and 600 $\mu mol \cdot L^{-1}$ for treatment. After the fruit were removed from the solution and air-dried at room temperature (22 ± 2 °C), they were placed into unsealed plastic bags (0.02 mm thick), and stored at 25 ± 1 °C with 70~80% relative humidity. Three independent biological replicates were performed for each treatment, where each replicate included 45 fruits. The fruit were periodically evaluated by ripening indicators. Samples were taken at 0, 3, 6, 8, 11, 13 and 15 days after treatment.

2.2. Fruit Firmness, Respiration Rate and Color Index

Fruit firmness was measured using an Instron Harness Tester 5542 (Instron, Norwood, NJ, USA) as described by Zhu et al. [31]. The measurements were conducted on five points of each fruit, and the results were expressed in Newtons.

The respiration rate was determined as described by Li et al. [32]. Briefly, the fruit were placed in a 2.5 L plastic container after weighing and held for 2 h at 25 °C. Next, 1 mL of headspace gas sample was extracted and measured using a gas chromatograph (G3900, Hitachi, Ltd., Tokyo, Japan). The respiration rate was expressed by the rates of CO_2 production, which were expressed as milligram per kilogram per hour.

Fruit color was measured as described by Li et al. [19] with minor modifications. Five points around the equatorial region on each fruit were selected for measurement using a Chromameter-2 reflectance colorimeter (Minolta, Osaka, Japan) equipped with a CR-300 measuring head. The results were recorded as lightness (L*), hue angle (h°), and chroma (C*).

2.3. Fruit Disease Evaluation

The incidence of disease and severity on the fruit were observed periodically as described by Pandey et al. [33] with minor modifications. The disease severity was recorded daily by measuring the lesion area on each fruit as compared to the total fruit surface area using a scale as follows: 0 = no lesion, 1 = 1–10%, 2 = 11–20%, 3 = 21–30%, 4 = 31–40%, 5 = 41–50%, 6 = 51–60%, 7 = 61–70%, 8 = 71–80%, 9 = 81–90%, and 10 = 91–100% of the area of the fruit was diseased.

The incidence of disease was calculated as the ratio of the number of diseased fruit to the total number of fruit [4].

2.4. Determination of Reactive Oxygen Species (ROS) and Malondialdehyde Contents

The contents of hydrogen peroxide (H_2O_2), superoxide radicals (O^{2-}) and malondialdehyde (MDA) were measured by UV-Vis spectrophotometry using kits (No. H2O2-2-Y, SA-2-G, and MDA-2-Y; Suzhou Keming Biological Company, Suzhou, China), according to the manufacturer's instructions. The absorbance was measured at 530 nm and calculated for the O^{2-} content with a unit expressed as nmol·g^{-1}. The absorbance at 415 nm was recorded for the content of H_2O_2 with a unit expressed as μmol·mL^{-1}. The content of MDA was determined by measuring the absorbance at 532 nm, and the unit was expressed as nmol·g^{-1}.

2.5. Enzyme Activities and Metabolites Content Measurement

2.5.1. The Antioxidative Enzyme Activities and Antioxidant Content Detection

The activities of catalase (CAT), superoxide dismutase (SOD), glutathione reductase (GR), ascorbate peroxidase (APX) and the total antioxidant capacity (T-AOC) were determined by UV-Vis spectrophotometry using different biochemical kits (No. SOD-2-W, CAT-2-W, GR-2-W, APX-2-W, ASA-1-W, and FRAP-2-G; Suzhou Keming Biological Company), according to the manufacturer's instructions. The absorbance from the activity of SOD was 450 nm, which was expressed as U·g^{-1}, and that of CAT was 240 nm, which was expressed as nmol/min^{-1}/g^{-1}. The absorbance value at 340 nm was recorded, and the unit was expressed as nmol·min^{-1}·g^{-1} for GR activity. For APX activity, the absorbance was recorded at 290 nm, and the unit was expressed as μmol·min^{-1}·g^{-1}. The absorbance was recorded at 593 nm for T-AOC and the unit expressed as μmol Trolox^{-1}·g^{-1}.

2.5.2. The Determination of Flavonoids and Ascorbic Acid Contents

The contents of total flavonoids and reduced ascorbic acid (ASA) were determined using a total flavonoid content kit and an ASA content kit (No. LHT-2-G, No. ASA-1-W; Suzhou Keming Biological Company) according to the manufacturer's instructions. The absorbance was measured at 510 nm and the unit expressed in mg·g^{-1} for the total flavonoid content. The content of ASA was measured at 420 nm and expressed as μg·g^{-1}.

2.5.3. Determination of the Activities of Enzymes Related to Phenylpropanoid Metabolism

The activities of three key enzymes involved in phenylpropanoid metabolism were determined, including 4-coumaric acid: coenzyme A ligase (4CL), cinnamic acid-4-hydroxylase activity (C4H) and phenylalanine ammonia lyase (PAL). The activities of 4CL and C4H were determined using an enzyme activity kit (No. G1003W and G1001W; Suzhou Geruisi Biological Company, Suzhou, China). For 4CL, the rate of production of 4-coumaric acid CoA at 333 nm was measured to determine the activity of 4CL in nmol·min^{-1}·g^{-1}. The absorbance was read at 340 nm for C4H and expressed as nmol·min^{-1}·g^{-1}.

The activity of PAL was measured at 290 nm and expressed as U·g^{-1} using a kit according to the manufacturer's instructions (No. PAL-2-Y; Suzhou Keming Biological Company).

2.5.4. Determination of Enzyme Activities Related to the Lipid Metabolism of Fruit Membranes

The activities of lipase (LPS), lipoxygenase (LOX) and phospholipase D (PLD) were determined using kits according to the manufacturer's instructions (No. G0902F, G0906F, and G0925F; Suzhou Geruisi Biotechnology Co. Ltd.). The activity of LPS was measured at 405 nm and expressed as μmol·min^{-1}·g^{-1}. The activity of LOX was assayed at 234 nm and expressed as U·g^{-1}. For PLD, the absorption was measured at 500 nm and expressed as nmol·min^{-1}·g^{-1}.

2.5.5. Determination of Enzyme Activities Related to Fruit Resistance

The activity of chitinase (CHI) activity was determined using a CHI content microassay kit (No. GO546W; Suzhou Geruisi Biological Company). The absorption at 585 nm was measured and expressed as mg·h^{-1}·g^{-1}.

The β-1,3-glucanase (β-1,3-GA) activity was assayed using a β-1,3-GA activity kit (No. GA-1-Y; Suzhou Keming Biological Company). The absorption at 550 nm was measured and the unit expressed as mg·h^{-1}·g^{-1}.

2.6. Statistical Analysis

The data were analyzed using SPSS 17.0 (SPSS Inc., Chicago, IL, USA) and Microsoft Excel 2019 (Redmond, WA, USA) to calculate the means and standard deviation (SD). The figures were plotted using SigmaPlot 12.0 (Systat, Chicago, IL, USA). Each experiment was conducted using three biological repetitions. Significant differences between treatment groups were determined using Duncan's multiple range test and least significant difference (LSD) ($p < 0.05$). All the results are expressed as the means ± SD (standard deviation) (Supplemental Table S1).

3. Results

3.1. The Effects of Melatonin on Fruit Ripening

Fruit firmness is one of the most important indicators for fruit ripening. As shown in Figure 1A, the fruit firmness decreased gradually with fruit storage under controlled conditions. All the melatonin treatments delayed the decrease in fruit firmness and maintained higher fruit firmness compared with the control (CK) group, particularly for the 600 μmol·L^{-1} treatment, which had significantly higher firmness than the other treatments at the end of storage (Figure 1A). The color of the peel was primarily evaluated by L*, C*, and h° values, which indicate the brightness, color saturation and the hue angle of the peel, respectively. As shown in Figure 1B, the L* value gradually increased with fruit ripening for the control group. No significant difference was observed between the melatonin treatment groups and the control group for the L* value (Figure 1B). The C* value of fruit increased with fruit ripening, and no significant difference was observed between the different groups (Figure 1C). The h° value gradually decreased with fruit ripening. The melatonin treatments delayed the decrease in h° value, which resulted in fruit with a higher h° value than the control during later storage (Figure 1D).

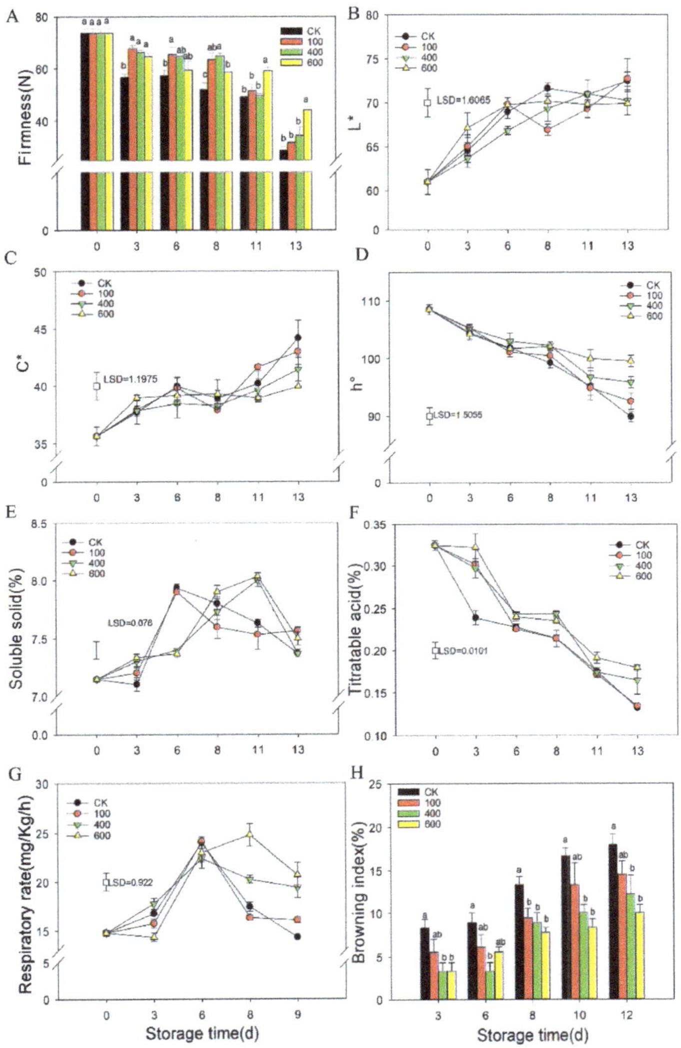

Figure 1. Effect of melatonin treatments on fruit physiology including the hardness, fruit color changes, soluble solid content, titratable acid, respiration rate and browning index. (**A**), the effect of melatonin treatment on the hardness of guava fruit; (**B–D**), the effect of melatonin treatment on the chromaticity L*, C*, h° of guava fruit; (**E**), the effect of melatonin treatment on soluble solid content; (**F**), the effect of melatonin treatment on the titratable acid content; (**G**), the effect of melatonin treatment on the respiration rate of guava fruit; (**H**), the effect of melatonin treatment on the browning index of guava fruit. Each data point represents the mean ± S.D. (n = 3). Different letters indicated the significant differences at the 5% level. Least significant differences (LSDs) were calculated to compare significant effects at the 5% level. Different letters indicate significant differences between groups ($p < 0.05$). CK: control group; 100: 100 $\mu mol \cdot L^{-1}$ melatonin treatment group; 400: 400 $\mu mol \cdot L^{-1}$ melatonin treatment group; 600: 600 $\mu mol \cdot L^{-1}$ melatonin treatment group.

The soluble solid content of the fruit increased with ripening and reached a peak at six days and then decreased gradually in the control group. No significant difference was observed between the 100 $\mu mol \cdot L^{-1}$ melatonin treatment group and the control group. However, the 400 $\mu mol \cdot L^{-1}$ and 600 $\mu mol \cdot L^{-1}$ melatonin treatments delayed the soluble solid content peak by approximately five days and maintained a higher content during later storage (Figure 1E). The titratable acid content decreased with fruit ripening, and the melatonin treatments reduced the decrease more than the control group (Figure 1F). Fruit respiration increased with ripening and peaked on day six before gradually decreasing (Figure 1F). The 600 $\mu mol \cdot L^{-1}$ melatonin treatment repressed the respiration rate at day three but enhanced it during the later storage. The 400 $\mu mol \cdot L^{-1}$ melatonin treatment also enhanced the respiration rate during the later storage with no difference during the first six days. No significant significance was observed between the 100 $\mu mol \cdot L^{-1}$ melatonin treatment and the control group during the entire storage period (Figure 1G). Fruit showed browning symptoms from day three of storage, and the browning index increased with fruit storage under the control conditions. The melatonin treatments reduced the browning index, particularly for the 600 $\mu mol \cdot L^{-1}$ melatonin treatment, except for the 100 $\mu mol \cdot L^{-1}$ melatonin treatment (Figure 1H).

3.2. *Effects of Different Concentrations of Melatonin on the Development of Fruit Disease*

Anthracnose is the most common disease for guava during the postharvest period. As shown in Figure 2, serious disease symptoms on the control fruit were observed on day 11 after harvest. Melatonin treatments significantly reduced and delayed the development of disease. This was particularly true for the 600 $\mu mol \cdot L^{-1}$ melatonin treatment in which disease was only observed after 15 days of storage (Figure 2A). As shown in Figure 2B, the disease index increased rapidly with fruit ripening. Melatonin treatments inhibited the development of disease index, which was significantly lower than that of the control fruit during the later storage, particularly for the 600 $\mu mol \cdot L^{-1}$ melatonin treatment (Figure 2B). Similar results were observed for the disease incidence (Figure 2C), and treatment with 600 $\mu mol \cdot L^{-1}$ melatonin resulted in the lowest disease incidence rate of all of the groups.

Additionally, the Principal Component Analysis (PCA) also identified that both concentrations and storage times of melatonin treatments have significant effects on the overall variation of physiological/biochemical features in guava fruit (Figure S1).

3.3. *Effects of Melatonin on the Metabolism of Reactive Oxygen Species*

Reactive oxygen species (ROS) are important in fruit ripening and stress responses. Hydrogen peroxide (H_2O_2) and O^{2-} are important ROS in plants when they are subjected to adverse conditions. As shown in Figure 3, the content of H_2O_2 gradually increased with fruit ripening and decreased at the end of storage in the control group. Treatment with melatonin reduced the content of H_2O_2 during the later storage resulting in significantly lower levels than those in the control group (Figure 3A). The content of O^{2-} gradually decreased as the fruit ripened and increased slightly at the end of storage in the control group. The melatonin treatment reduced the content of O^{2-} during the entire storage period (Figure 3B). Various antioxidants and antioxidant enzymes in guava fruit constitute the total antioxidant capacity (T-AOC). The T-AOC capacity decreased gradually with fruit storage, and treatment with melatonin maintained a high capacity of T-AOC during the first eight days of storage. However, there was no significant difference compared with the control group during the later storage period (Figure 3C). MDA is the final product of cell lipid peroxidation. As shown in Figure 3D, the content of MDA increased gradually in the control group as the fruit ripened (Figure 3C). Treatment with melatonin significantly reduced the content of MDA during the whole storage period, which was significantly lower than that of the control group, indicating a reduced degree of the level of peroxidation of guava fruit.

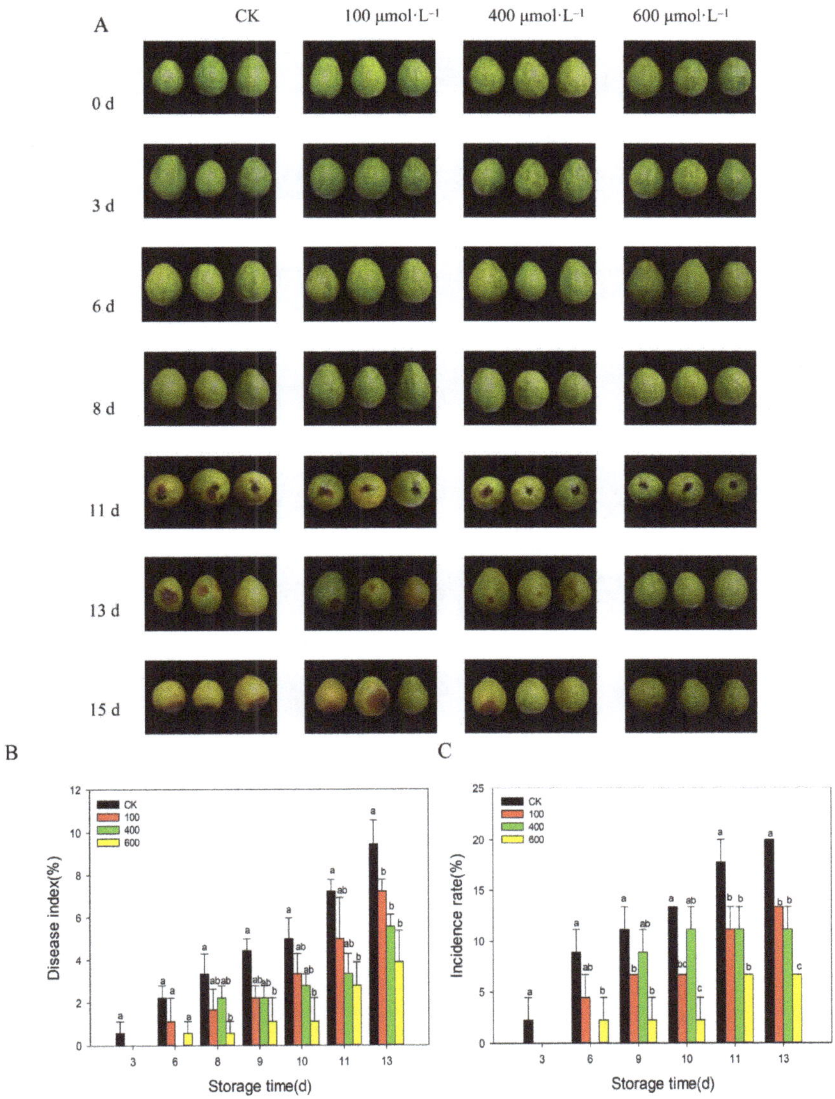

Figure 2. Effects of different concentrations of melatonin on fruit disease incidence. (**A**) Photos of guava fruit under different melatonin treatments; (**B**) fruit disease index; (**C**) fruit disease incidence. Guava fruit were stored at room temperature (25 ± 1 °C) after treatment. Each data point represents the mean ± S.D. (n = 3). Different letters indicated the significant differences at the 5% level. CK: control group; 100: 100 μmol·L^{-1} melatonin treatment group; 400: 400 μmol·L^{-1} melatonin treatment group; 600: 600 μmol·L^{-1} melatonin treatment group.

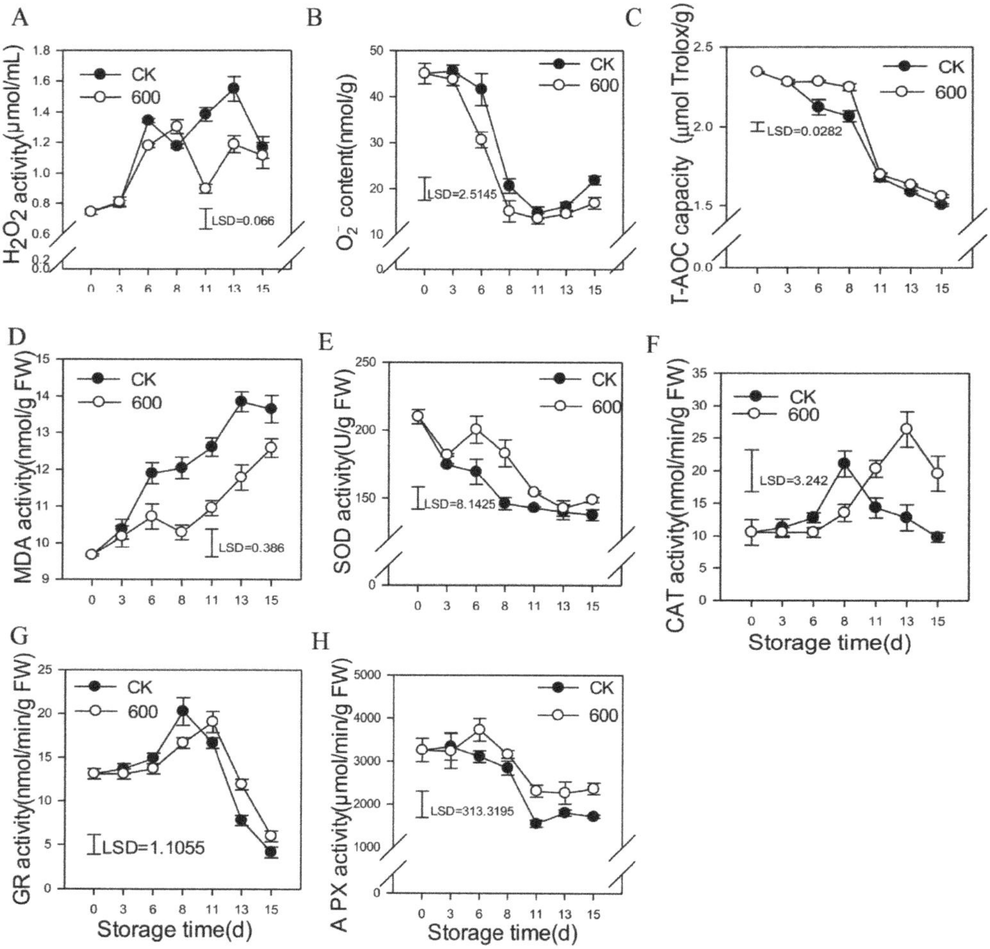

Figure 3. Effects of melatonin on the metabolism of reactive oxygen species in guava fruit. (**A**) the effect of melatonin treatment on H_2O_2 content; (**B**) the effect of melatonin treatment on O^{2-} content. (**C**) the effect of melatonin treatment on T-AOC capacity; (**D**) the effect of melatonin treatment on the MDA content; (**E–H**) the effect of melatonin treatment on the activities of SOD (**E**), CAT (**F**), GR (**G**) and APX (**H**). Each data point represents the mean ± S.D. (n = 3). Least significant differences (LSDs) were calculated to compare significant effects at the 5% level. CK: control group; 600: 600 $\mu mol \cdot L^{-1}$ melatonin treatment group.

Different active oxygen scavengers function to maintain the balance of ROS metabolism and protect the membrane structure. SOD is an important antioxidant enzyme that is widespread in various organisms and catalyzes the dismutation of O^{2-} to hydrogen and molecular oxygen. CAT is a reactive oxygen scavenging enzyme that can degrade H_2O_2 into O_2 and H_2O to reduce the damage to plant tissues. GR is a flavoprotein oxidoreductase that is ubiquitous in eukaryotes and prokaryotes, and it is one of the important antioxidant enzymes for the active oxygen metabolism in plants. APX has some ability to remove active oxygen, which can effectively remove the H_2O_2 content in plants and effectively enhance the resistance of plants to external stresses.

As shown in Figure 3E, the activity of SOD decreased with fruit storage, and treatment with melatonin induced its activity, which was significantly higher than that of the control fruit. The activity of CAT increased first and reached a peak at day eight before gradually decreasing. Treatment with melatonin delayed the increase, resulting in much higher levels of activity compared with the control group (Figure 3F). Treatment with melatonin delayed the increase in the activity of GR during earlier storage and decreased during later storage, thus, maintaining higher GR activity during this period (Figure 3G). Melatonin also enhanced the activity of APX, which was higher compared with the control fruit during storage (Figure 3H).

3.4. Effects of Melatonin Treatment on the Content of Defense-Related Compounds

Flavonoids are important secondary metabolites in plants. They have strong antioxidant activity and affect the color and flavor of the fruit. Flavonoids are a class of active substances that act as antioxidants and scavenge free radicals. As shown in Figure 4, the content of total flavonoids decreased with fruit storage, and treatment with melatonin reduced the decrease and maintained higher flavonoid contents than the control fruit during the later period of storage (Figure 4A).

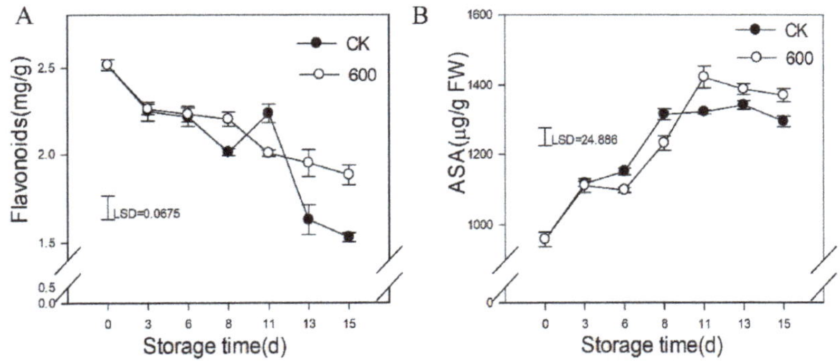

Figure 4. Effects of melatonin treatment on the content of defense-related substance in guava fruit. (**A**) the effect of melatonin treatment on flavonoids content; (**B**) the effect of melatonin treatment on ASA content. Each data point represents the mean ± S.D. (n = 3). Least significant differences (LSDs) were calculated to compare significant effects at the 5% level.CK: control group; 600: 600 $\mu mol \cdot L^{-1}$ melatonin treatment group.

The content of ASA increased with fruit ripening, and treatment with melatonin induced the accumulation of ASA during the later period of storage (Figure 4B).

3.5. Effects of Melatonin Treatment on the Activities of Enzymes in the Phenylpropanoid Pathway

C4H is a key enzyme in the phenylpropanoid metabolic pathway, which can effectively regulate the synthesis of corresponding flavonoids and other metabolites. As shown in Figure 5A, the C4H activity generally decreased with fruit ripening and increased from days six to eleven. However, it decreased slightly during the later storage period (Figure 5A). Treatment with melatonin enhanced the activity of C4H on day three but reduced its activity during later storage (Figure 4A). Phenylalanine ammonia lyase (PAL) is one of the rate-limiting enzymes that plays critical roles in plant resistance. The activity of PAL increased gradually as the fruit ripened, and treatment with melatonin induced its activity. Significantly higher levels of PAL were observed in fruit treated with melatonin than the control during the later storage (Figure 5 B). 4CL is one of the key enzymes in lignin synthesis and catalyzes cinnamic acid to produce corresponding cinnamic acid-CoA esters, which are primarily found in higher plants, yeasts and fungi. The activity of 4CL had a trend of fluctuating–decreasing–increasing–decreasing–increasing. Treatment with melatonin induced its activity at some level (Figure 5C).

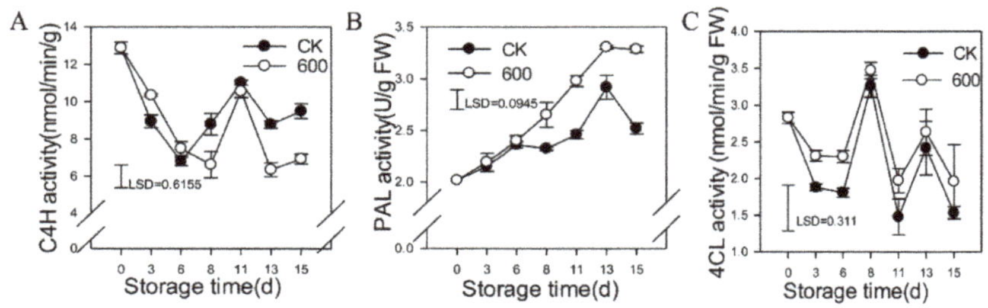

Figure 5. Effects of melatonin treatment on enzyme activities of the phenylpropanoid pathway in guava fruit. (**A–C**), the effect of melatonin treatment on the activities of C4H (**A**); PAL (**B**) and 4CL (**C**). Each data point represents the mean ± S.D. (n = 3). Least significant differences (LSDs) were calculated to compare significant effects at the 5% level. CK: control group; 600: 600 $\mu mol \cdot L^{-1}$ melatonin treatment group.

3.6. Effects of Melatonin on the Activities of Lipid Metabolic Enzymes

By catalyzing the oxidation reaction of unsaturated fatty acids, LOX can lead to membrane lipid peroxidation and play an important role in the maturation, senescence and adversity of stress in organisms. The activity of LOX increased gradually as the fruit ripened and reached its peak at day eight before decreasing. Treatment with melatonin reduced the activity of LOX, which was significantly lower than that of the control fruit during the first 11 days of storage (Figure 6A). LPS, also known as glyceride hydrolase, is widely present in a variety of organisms and can catalyzes the hydrolysis of triglycerides to produce fatty acids and glycerol. As shown in Figure 6B, the activity of LPS increased and reached a peak at day six, decreased rapidly and then increased slightly during the later period of storage. Treatment with melatonin severely repressed the activity of LPS during this period of storage (Figure 6B). The activity of PLD increased as the fruit ripened, and treatment with melatonin induced its activity during the first six days but severely inhibited its activity during later storage (Figure 6C).

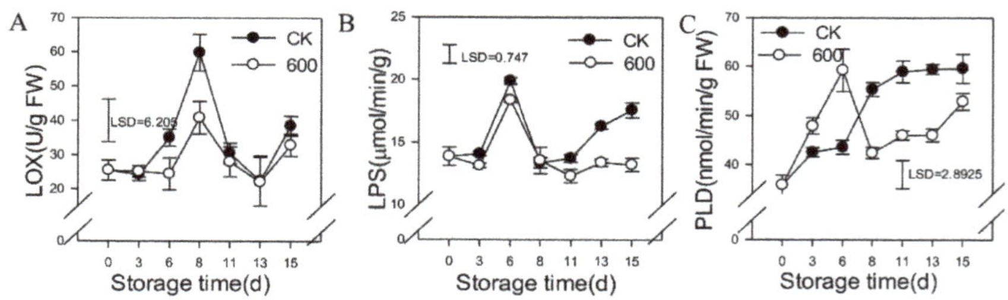

Figure 6. Effect of melatonin treatment on the activities of lipid metabolism enzymes in guava fruit. (**A–C**), the effect of melatonin treatment on the activities of LOX (**A**); LPS (**B**); PLD (**C**). Each data point represents the mean ± S.D. (n = 3). Least significant differences (LSDs) were calculated to compare significant effects at the 5% level. CK: control group; 600: 600 $\mu mol \cdot L^{-1}$ melatonin treatment group.

3.7. Effects of Melatonin Treatment on the Activities of Enzymes Related to Resistance

CHI is a defense-related enzyme that is found widely in animals and plants. As shown in Figure 7A, the activity of CHI increased with fruit ripening and reached its peak on day 11 and then decreased during later storage (Figure 7A). Treatment with melatonin induced the activity of CHI, which had higher activity compared with the control group. The activity of β-1,3-GA increased gradually as the fruit ripened. Melatonin repressed the activity of β-1,3-GA during storage (Figure 7B).

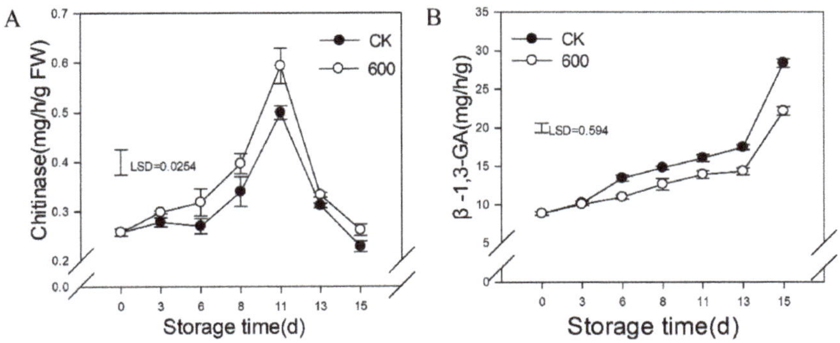

Figure 7. Effect of melatonin treatment on the activities resistance related enzymes in guava fruit. (**A**,**B**) the effect of melatonin treatment on the activities of CHI (**A**); β-1,3-GA (**B**). Each data point represents the mean ± S.D. (n = 3). Least significant differences (LSDs) were calculated to compare significant effects at the 5% level. CK: control group; 600: 600 μmol·L^{-1} melatonin treatment group.

4. Discussion

Melatonin is an important signaling molecule that functions as a master regulator in various developmental stages of plants and their adaptation to the environment [34,35]. Increasing numbers of studies have proven that the application of exogenous melatonin can effectively prolong storage and shelf life and maintain fruit quality [23–27]. A few studies also showed that the exogenous application of melatonin could accelerate the ripening of fruit, such as tomatoes. Treatment with melatonin substantially enhanced the accumulation of the content of lycopene and the production of ethylene in tomatoes, resulting in accelerated fruit ripening and an improvement in the quality of appearance of the tomato fruit [18]. Exogenous melatonin treatment on young grapes also significantly promoted the growth and increased the endogenous melatonin content of the fruit, significantly increasing the content of total anthocyanins, fruit total soluble solids and promoting the ripening of grapes [36]. This study demonstrates that treatment with melatonin could effectively delay the softening of guava fruit, which is consistent with most of the studies so far. Additionally, treatment with melatonin significantly reduces the incidence of anthracnose in guava fruit. Other studies also showed that the exogenous application of melatonin can significantly enhance the disease resistance and reduces the occurrence of decay of different fruit during storage by improving the antioxidant and defense systems of the fruit, such as peaches [26], strawberries [29], kiwifruits [30] and plums [37]. Therefore, as a safe and nontoxic substance, melatonin has a practical potential for application in postharvest guava fruit. Actually, the effects of melatonin are dose-dependent [35]. It was reported that 100 μM melatonin treatment could promote the ripening of grape fruit, while 0.1 and 1.0 μM melatonin showed no significant effect [38]. Melatonin application before flowering could delay the flowering of apple trees in a dose-dependent manner. The increased melatonin levels at a suitable range also resulted in more flowering, but unsuitably high concentrations would repress flowering [16]. The present work also showed that the effect of melatonin on fruit ripening is dose-dependent. Hence, the optimized concentration should be identified during application.

Fruit ripening often concomitantly proceeds with the accumulation of ROS, which function as the accelerator of the ripening process by adversely oxidizing membrane lipids, structural proteins and nucleic acids [39]. Fruit have evolved efficient non-enzymatic and enzymatic antioxidative systems to scavenge ROS and protect against oxidative damage [40]. The antioxidant enzymes, including CAT, APX, polyphenol oxidase (PPO), peroxidase (POD), dehydroascorbate reductase, SOD, GR and GPX, are important for ROS homeostasis in fruit [14]. The non-enzymatic antioxidants include ascorbic acid (AsA), flavonoids, anthocyanins, phenolics, carotenoids, dehydroascorbic acid and glutathione (GSH) [14,41]. Melatonin works as an excellent antioxidant molecule in plants. The application of exogenous melatonin can effectively reduce the accumulation of ROS in different fruits by inducing antioxidant systems, including the enzymatic antioxidants and non-enzymatic antioxidants, and then delay the process of fruit ripening [22,42,43]. For example, treatment with melatonin significantly enhanced the antioxidant capacity by increasing the total phenolic content, the scavenging ability of 1,1-diphenyl-2-picrylhydrazyl (DPPH) and 2,2′-Azinobis(3-ethylbenzothiazoline-6-sulfonic acid Ammonium Salt) (ABTS) and the activities of PAL and CHS, reducing the contents of MDA and H_2O_2 and decreasing ascorbic acid in fresh-cut pear fruit, resulting in reduced browning and microbial growth [44]. In strawberry fruit, treatment with exogenous melatonin significantly reduced the contents of H_2O_2 and MDA but enhanced the accumulation of total phenolics and flavonoids, resulting in an improved antioxidant capacity and a delay in fruit ripening [45]. In peach fruit, treatment with melatonin effectively increased the activities of CAT, SOD, APX and POD and reduced the content of MDA, O^{2-} and hydrogen peroxide, maintained the integrity of membranes and resulted in delayed postharvest senescence [26]. The application of melatonin also significantly increased the activities of CAT, SOD and POD in apple fruits, improving the antioxidant capacity of the fruit and maintaining its quality [22]. Non-enzymatic antioxidants, such as phenolics, GSH, AsA, flavonoids and anthocyanins, were also effectively induced by melatonin treatment [14,35,37]. The exogenous application of melatonin increased the polyphenol content and antioxidant activities in grapes [46]. This treatment also maintained the nutritional quality and reduced the decay of postharvest strawberry fruit by enhancing the contents of total phenolics, the accumulation of anthocyanins and γ-aminobutyric acid transaminase (GABA) shunt activity [29]. The exogenous application of melatonin increased the amount of phenolic compounds by upregulating the expression of related genes, which led to a delay in fruit senescence and deterioration in the quality of jujube fruit [47]. In this study, melatonin also effectively enhanced enzymatic antioxidants, such as the activities of SOD, GR, CAT and APX, and the content of non-enzymatic antioxidants, such as the total flavonoids, TP and ASA, and improving the T-AOC of the fruit, resulting in a reduction in the content of H_2O_2, O^{2-} and MDA. All these results proved that melatonin serves as an effective ROS scavenger to protect fruit from oxidative damage and maintain their quality.

Postharvest decays caused by diseases are a major cause of losses during the postharvest period. Most postharvest diseases could be controlled effectively by chemical fungicide treatments, but their use is strictly restricted owing to the increased enhancement of food safety awareness. Melatonin, a safe and nontoxic substance to humans and plants, functions as an environmentally friendly treatment that may be an alternative to fungicide applications and has shown its potential in postharvest preservation. A large number of studies have proven that melatonin can effectively reduce the postharvest disease incidence of various fruit, including bananas, apples, litchis, grapes, peaches, strawberries, plums and tomatoes [14,28]. The application of exogenous melatonin induced resistance to *Peronophythora litchii* in litchis by enhancing the activities of enzyme antioxidants, including PAL, C4H and 4CL, and increasing the accumulation of non-enzyme antioxidants, including phenolics and flavonoids [28]. Melatonin combined with *Meyerozyma guilliermondii* Y-1 effectively enhanced defense-related enzyme activities, including SOD, CAT, POD, PAL, and PPO, as well as increasing the content of non-enzyme antioxidants, such as the contents of TP and lignin, and the total antioxidant capacity (T-AOC), resulting in reduced disease incidence in apple fruit [48]. This study also showed that treatment with melatonin effectively reduces the incidence of anthracnose on fruit, which may result from an enhancement of the enzymatic antioxidants and non-enzymatic antioxidants, and the enhanced phenylpropanoid pathways, as well as an enhancement of the defense-related enzyme activities, including CHI. Some studies have shown that melatonin exhibited antimicrobial activities against different fruit pathogens in vitro and in vivo. In tomatoes, melatonin treatment effectively suppressed the growth of *Phytophthora infestans* in vitro and reduced the late blight symptoms of tomatoes in vivo [49]. Exogenous treatment with melatonin also induced the accumulation of arecoline and modulated the contents of ROS and levels of hormones, including ABA, JA, ethylene, and SA, as well as glycolytic activity in areca (*Arecha catechu*) fruit, resulting in improved plant disease resistance against *C. kahawae* and delayed postharvest physiological deterioration [50]. Melatonin can effectively induce the phenylpropanoid pathway and enhance the intermediate pathway for the synthesis of disease resistance-related phenolics in various fruit species. For example, in cherry tomato fruit, the application of melatonin effectively enhanced the accumulation of total phenolics, flavonoids and lignin by increasing the enzyme activities of the phenylpropanoid pathway, including 4CL, PAL and POD [21]. In tomatoes, melatonin treatment effectively reduced the development of the development of the virulent pathogen *Botrytis cinerea* by increasing the amounts of the endogenous resistance hormones SA, methyl jasmonate (MeJA) and melatonin, inducing an ROS burst, and enhancing the activities of the resistance enzymes of CHI and GLU and the phenylpropanoid pathway [21,51]. Melatonin treatment can also reduce the amount of pathogen infection by inducing the synthesis of cell walls, lipids and waxes in banana peels [19]. However, a few reports showed that melatonin could enhance disease. For example, melatonin treatment decreases the resistance of citrus fruit to postharvest green mold by reducing the level of defense-related ROS [52]. Strikingly, the effect of melatonin on fruit quality and postharvest disease could be owing to the signal-crossing talk in fruit and vegetables. There is widespread signal-crossing talk after melatonin treatment. Melatonin affects the signal molecules, such as ethylene, IAA, ABA, SA, GA, and NO among others, as well as endogenous melatonin in plants, to regulate fruit ripening and stress responses [14]. The application of exogenous melatonin can also induce the accumulation of endogenous melatonin during the fruit ripening process and affect fruit quality and shelf life of such fruit as grapes [53], pears [25], sweet cherries [23] and bananas [24]. Thus, the effects of melatonin treatment can be dose-dependent or species-dependent.

Fruit respiration is important for fruit physiology during ripening process. Melatonin treatment effectively reduced the elevation of the respiratory rate in bananas [19], pears [25], mango fruits [27], as well as altered the secondary metabolism pathways, resulting in the delayed ripening process. Preharvest spray with melatonin depressed fruit respiration after harvest while melatonin induced respiratory activity following cold storage [54]. It indicated that the sweet cherry response to melatonin treatment involved a cold-dependent activation of the respiration and up-regulation of TCA cycle-related genes [54]. It seems that cold storage may cause stress to sweet cherry fruit and give rise to ROS. Melatonin may enhance the respiration and improve the energy supply to maintain fruit quality [28]. Importantly, melatonin induced phenolic compound accumulation and related genes expression, which help to improve fruit quality and cold tolerance [54]. In the present work, it was established that melatonin repressed fruit respiration rates during the early storage period, but increased respiration during later storage, which corresponded to the enhanced total flavonoids and ascorbic acid accumulation. The increased respiration may contribute to the accumulation of secondary metabolites, helping the disease response. Melatonin treatment also maintained fruit firmness and reduced the change in fruit color, retarding fruit ripening and senescence.

5. Conclusions

As shown in Figure 8, in this study we showed that treatment with melatonin effectively enhanced the antioxidant capacity, the content of total flavonoids and ascorbic acid and the activity of phenylpropanoid pathway enzymes, such as PAL and 4CL, and reduced the oxidative damage of guava fruit. Melatonin treatment also enhanced the activities of defense-related enzymes, including CHI, and repressed the activities of LPS, LOX and PLD related to lipid metabolism. All these results showed that treatment with exogenous melatonin can maintain fruit quality and enhance disease resistance by improving the antioxidant and defense systems.

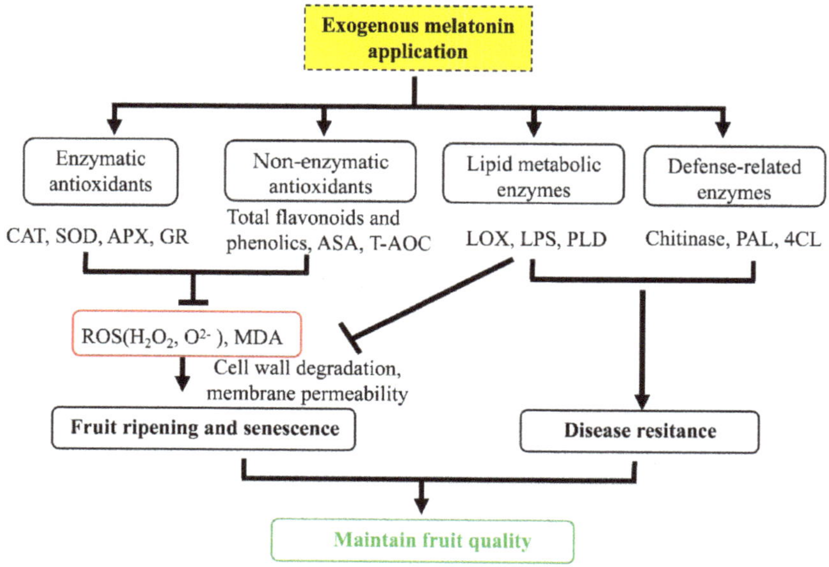

Figure 8. Proposed roles of exogenous melatonin treatment to maintain postharvest guava fruit quality. ROS: reactive oxygen species; MDA: malondialdehyde; CAT: catalase; SOD: superoxide dismutase; GR: glutathione reductase; APX: ascorbate peroxidase; ASA: ascorbic acid; 4CL: 4-coumaric acid: coenzyme A ligase; PAL: phenylalanine ammonia lyase; LPS: lipase; LOX: lipoxygenase; PLD: phospholipase D.

Supplementary Materials: The following supporting information can be downloaded at: https://www.mdpi.com/article/10.3390/foods11030262/s1, Figure S1: Overview of physiological index changes presented in Figures 1 and 2 of guava fruit in response to melatonin treatments during storage; Table S1: Dataset for all figures in present work.

Author Contributions: X.Z. and X.L. designed the experiments; S.F., Q.L., Q.T. and J.C. conducted the experiments; X.Z., T.X., Z.S., M.Y. and S.F. interpretated the data, X.Z. and T.X. wrote the manuscript. W.C., M.Y. and X.L. reviewed the manuscript. All authors have read and agreed to the published version of the manuscript.

Funding: This research was funded by National Natural Science Foundation of China (grants no. 32172640), Natural Science Foundation of Guangdong Province (grants no. 2021A1515012435), Guangzhou Science and Technology Project Scientific Special (No201904010011), Pearl River Talent Program for Young Talent (grant no. 2017GC010321) and the Science and Technology Planning Project of Guangdong Province (2018A050506074),Young Innovative Talents Projects in Ordinary Colleges and Universities in Guangdong Province (grant no. 2021KQNCX005).

Data Availability Statement: The data presented in this study are available on request from the corresponding author.

Conflicts of Interest: The authors declare that the research was conducted in the absence of any commercial or financial relationships that could be construed as a potential conflict of interest.

References

1. Singh, S.P. Guava (Psidium guajava L.). In *Postharvest Biology and Technology of Tropical and Subtropical Fruits: Cocona to Mango*; Woodhead Publishing: Sawston, UK, 2011; pp. 213–246e. ISBN 9781845697358.
2. McCook-Russell, K.P.; Nair, M.G.; Facey, P.C.; Bowen-Forbes, C.S. Nutritional and nutraceutical comparison of Jamaican Psidium cattleianum (strawberry guava) and Psidium guajava (common guava) fruits. *Food Chem.* **2012**, *134*, 1069–1073. [CrossRef] [PubMed]

3. Murmu, S.B.; Mishra, H.N. Post-harvest shelf-life of banana and guava: Mechanisms of common degradation problems and emerging counteracting strategies. *Innov. Food Sci. Emerg. Technol.* **2018**, *49*, 20–30. [CrossRef]
4. Singh, S.P.; Pal, R.K. Response of climacteric-type guava (Psidium guajava L.) to postharvest treatment with 1-MCP. *Postharvest Biol. Technol.* **2008**, *47*, 307–314. [CrossRef]
5. González-Aguilar, G.A.; Tiznado-Hernández, M.E.; Zavaleta-Gatica, R.; Martínez-Téllez, M.A. Methyl jasmonate treatments reduce chilling injury and activate the defense response of guava fruits. *Biochem. Biophys. Res. Commun.* **2004**, *313*, 694–701. [CrossRef]
6. Zakaria, L. Diversity of Colletotrichum Species Associated with Anthracnose Disease in Tropical Fruit Crops—A Review. *Agriculture* **2021**, *11*, 297. [CrossRef]
7. Oliveira, P.D.L.; de Oliveira, K.Á.R.; Dos Santos Vieira, W.A.; Câmara, M.P.S.; de Souza, E.L. Control of anthracnose caused by *Colletotrichum* species in guava, mango and papaya using synergistic combinations of chitosan and *Cymbopogon citratus* (D.C. ex Nees) Stapf. essential oil. *Int. J. Food Microbiol.* **2018**, *266*, 87–94. [CrossRef]
8. Alba-Jiménez, J.E.; Benito-Bautista, P.; Nava, G.M.; Rivera-Pastrana, D.M.; Vázquez-Barrios, M.E.; Mercado-Silva, E.M. Chilling injury is associated with changes in microsomal membrane lipids in guava fruit (*Psidium guajava* L.) and the use of controlled atmospheres reduce these effects. *Sci. Hortic.* **2018**, *240*, 94–101. [CrossRef]
9. Vishwasrao, C.; Ananthanarayan, L. Postharvest shelf-life extension of pink guavas (*Psidium guajava* L.) using HPMC-based edible surface coatings. *J. Food Sci. Technol.* **2016**, *53*, 1966–1974. [CrossRef]
10. Lo'ay, A.A.; Taher, M.A. Influence of edible coatings chitosan/PVP blending with salicylic acid on biochemical fruit skin browning incidence and shelf life of guava fruits cv. 'Banati.'. *Sci. Hortic.* **2018**, *235*, 424–436. [CrossRef]
11. Azam, M.; Hameed, L.; Qadri, R.; Ejaz, S.; Ghani, M.A. Postharvest ascorbic acid application maintained physiological and antioxidant responses of Guava (*Psidium guajava* L.) at ambient storage. *Food Sci. Technol.* **2021**, *40*, 748–754. [CrossRef]
12. Sahu, S.K.; Barman, K.; Singh, A.K. Nitric oxide application for postharvest quality retention of guava fruits. *Acta Physiol. Plant.* **2020**, *42*, 156. [CrossRef]
13. Botelho, R.V.; Souza, N.; Peres, N. Effect of the postharvest treatment with calcium chloride by the temperature differential method on the control of Colletotrichum gloeosporioides in guavas "Branca de Kumagaii". *Summa Phytopathol.* **2000**, *26*, 268–271.
14. Ze, Y.; Gao, H.; Li, T.; Yang, B.; Jiang, Y. Insights into the roles of melatonin in maintaining quality and extending shelf life of postharvest fruits. *Trends Food Sci. Technol.* **2021**, *109*, 569–578. [CrossRef]
15. Zhang, N.; Zhang, H.-J.; Sun, Q.-Q.; Cao, Y.-Y.; Li, X.; Zhao, B.; Wu, P.; Guo, Y.-D. Proteomic analysis reveals a role of melatonin in promoting cucumber seed germination under high salinity by regulating energy production. *Sci. Rep.* **2017**, *7*, 503. [CrossRef]
16. Zhang, H.; Wang, L.; Shi, K.; Shan, D.; Zhu, Y.; Wang, C.; Bai, Y.; Yan, T.; Zheng, X.; Kong, J. Apple tree flowering is mediated by low level of melatonin under the regulation of seasonal light signal. *J. Pineal Res.* **2019**, *66*, e12551. [CrossRef]
17. Wang, C.; Yin, L.-Y.; Shi, X.-Y.; Xiao, H.; Kang, K.; Liu, X.-Y.; Zhan, J.-C.; Huang, W.-D. Effect of Cultivar, Temperature, and Environmental Conditions on the Dynamic Change of Melatonin in Mulberry Fruit Development and Wine Fermentation. *J. Food Sci.* **2016**, *81*, M958–M967. [CrossRef]
18. Sun, Q.; Zhang, N.; Wang, J.; Zhang, H.; Li, D.; Shi, J.; Li, R.; Weeda, S.; Zhao, B.; Ren, S.; et al. Melatonin promotes ripening and improves quality of tomato fruit during postharvest life. *J. Exp. Bot.* **2015**, *66*, 657–668. [CrossRef]
19. Li, T.; Wu, Q.; Zhu, H.; Zhou, Y.; Jiang, Y.; Gao, H.; Yun, Z. Comparative transcriptomic and metabolic analysis reveals the effect of melatonin on delaying anthracnose incidence upon postharvest banana fruit peel. *BMC Plant Biol.* **2019**, *19*, 289. [CrossRef]
20. Cao, S.; Bian, K.; Shi, L.; Chung, H.-H.; Chen, W.; Yang, Z. Role of Melatonin in Cell-Wall Disassembly and Chilling Tolerance in Cold-Stored Peach Fruit. *J. Agric. Food Chem.* **2018**, *66*, 5663–5670. [CrossRef]
21. Li, S.; Xu, Y.; Bi, Y.; Zhang, B.; Shen, S.; Jiang, T.; Zheng, X. Melatonin treatment inhibits gray mold and induces disease resistance in cherry tomato fruit during postharvest. *Postharvest Biol. Technol.* **2019**, *157*, 110962. [CrossRef]
22. Onik, J.C.; Wai, S.C.; Li, A.; Lin, Q.; Sun, Q.; Wang, Z.; Duan, Y. Melatonin treatment reduces ethylene production and maintains fruit quality in apple during postharvest storage. *Food Chem.* **2021**, *337*, 127753. [CrossRef] [PubMed]
23. Tijero, V.; Muñoz, P.; Munné-Bosch, S. Melatonin as an inhibitor of sweet cherries ripening in orchard trees. *Plant Physiol. Biochem.* **2019**, *140*, 88–95. [CrossRef] [PubMed]
24. Hu, W.; Yang, H.; Tie, W.; Yan, Y.; Ding, Z.; Liu, Y.; Wu, C.; Wang, J.; Reiter, R.J.; Tan, D.-X.; et al. Natural Variation in Banana Varieties Highlights the Role of Melatonin in Postharvest Ripening and Quality. *J. Agric. Food Chem.* **2017**, *65*, 9987–9994. [CrossRef]
25. Liu, J.; Yang, J.; Zhang, H.; Cong, L.; Zhai, R.; Yang, C.; Wang, Z.; Ma, F.; Xu, L. Melatonin Inhibits Ethylene Synthesis via Nitric Oxide Regulation To Delay Postharvest Senescence in Pears. *J. Agric. Food Chem.* **2019**, *67*, 2279–2288. [CrossRef] [PubMed]
26. Gao, H.; Zhang, Z.K.; Chai, H.K.; Cheng, N.; Yang, Y.; Wang, D.N.; Yang, T.; Cao, W. Melatonin treatment delays postharvest senescence and regulates reactive oxygen species metabolism in peach fruit. *Postharvest Biol. Technol.* **2016**, *118*, 103–110. [CrossRef]
27. Liu, S.; Huang, H.; Huber, D.J.; Pan, Y.; Shi, X.; Zhang, Z. Delay of ripening and softening in 'Guifei' mango fruit by postharvest application of melatonin. *Postharvest Biol. Technol.* **2020**, *163*, 111136. [CrossRef]
28. Zhang, Z.; Wang, T.; Liu, G.; Hu, M.; Yun, Z.; Duan, X.; Cai, K.; Jiang, G. Inhibition of downy blight and enhancement of resistance in litchi fruit by postharvest application of melatonin. *Food Chem.* **2021**, *347*, 129009. [CrossRef]

29. Aghdam, M.S.; Fard, J.R. Melatonin treatment attenuates postharvest decay and maintains nutritional quality of strawberry fruits (*Fragaria×anannasa* cv. Selva) by enhancing GABA shunt activity. *Food Chem.* **2017**, *221*, 1650–1657. [CrossRef] [PubMed]
30. Hu, M.; Li, J.; Rao, J. Effect of Melatonin on Ripening and Senescence of Postharvest Kiwifruits. *Food Sci.* **2018**, *39*, 226–232. [CrossRef]
31. Zhu, X.; Ye, L.; Ding, X.; Gao, Q.; Xiao, S.; Tan, Q.; Huang, J.; Chen, W.; Li, X. Transcriptomic analysis reveals key factors in fruit ripening and rubbery texture caused by 1-MCP in papaya. *BMC Plant Biol.* **2019**, *19*, 309. [CrossRef]
32. Li, X.; Zhu, X.; Mao, J.; Zou, Y.; Fu, D.; Chen, W.; Lu, W. Isolation and characterization of ethylene response factor family genes during development, ethylene regulation and stress treatments in papaya fruit. *Plant Physiol. Biochem.* **2013**, *70*, 81–92. [CrossRef] [PubMed]
33. Pandey, R.R.; Arora, D.K.; Dubey, R.C. Effect of environmental conditions and inoculum density on infection of guava fruits by Colletotrichum glososporioides. *Mycopathologia* **1997**, *137*, 165–172. [CrossRef]
34. Arnao, M.B.; Hernández-Ruiz, J. Melatonin: A New Plant Hormone and/or a Plant Master Regulator? *Trends Plant Sci.* **2019**, *24*, 38–48. [CrossRef] [PubMed]
35. Wu, X.; Ren, J.; Huang, X.; Zheng, X.; Tian, Y.; Shi, L.; Dong, P.; Li, Z. Melatonin: Biosynthesis, content, and function in horticultural plants and potential application. *Sci. Hortic.* **2021**, *288*, 110392. [CrossRef]
36. Yang, M.; Wang, L.; Belwal, T.; Zhang, X.; Lu, H.; Chen, C.; Li, L. Exogenous Melatonin and Abscisic Acid Expedite the Flavonoids Biosynthesis in Grape Berry of Vitis vinifera cv. *Kyoho. Molecules* **2020**, *25*, 12. [CrossRef] [PubMed]
37. Bal, E. Physicochemical changes in 'Santa Rosa' plum fruit treated with melatonin during cold storage. *J. Food Meas. Charact.* **2019**, *13*, 1713–1720. [CrossRef]
38. Xu, L.; Yue, Q.; Xiang, G.; Bian, F.; Yao, Y. Melatonin promotes ripening of grape berry via increasing the levels of ABA, H_2O_2, and particularly ethylene. *Hortic. Res.* **2018**, *5*, 41. [CrossRef]
39. Corpas, F.J.; Freschi, L.; Rodríguez-Ruiz, M.; Mioto, P.T.; González-Gordo, S.; Palma, J.M. Nitro-oxidative metabolism during fruit ripening. *J. Exp. Bot.* **2018**, *69*, 3449–3463. [CrossRef]
40. Jiang, Y.; Duan, X.; Joyce, D.; Zhang, Z.; Li, J. Advances in understanding of enzymatic browning in harvested litchi fruit. *Food Chem.* **2004**, *88*, 443–446. [CrossRef]
41. Mirshekari, A.; Madani, B.; Yahia, E.M.; Golding, J.B.; Vand, S.H. Postharvest melatonin treatment reduces chilling injury in sapota fruit. *J. Sci. Food Agric.* **2020**, *100*, 1897–1903. [CrossRef]
42. Wang, S.-Y.; Shi, X.-C.; Wang, R.; Wang, H.-L.; Liu, F.; Laborda, P. Melatonin in fruit production and postharvest preservation: A review. *Food Chem.* **2020**, *320*, 126642. [CrossRef] [PubMed]
43. Liu, J.; Liu, H.; Wu, T.; Zhai, R.; Yang, C.; Wang, Z.; Ma, F.; Xu, L. Effects of Melatonin Treatment of Postharvest Pear Fruit on Aromatic Volatile Biosynthesis. *Molecules* **2019**, *24*, 4233. [CrossRef]
44. Zheng, H.; Liu, W.; Liu, S.; Liu, C.; Zheng, L. Effects of melatonin treatment on the enzymatic browning and nutritional quality of fresh-cut pear fruit. *Food Chem.* **2019**, *299*, 125116. [CrossRef]
45. Liu, C.; Zheng, H.; Sheng, K.; Liu, W.; Zheng, L. Effects of melatonin treatment on the postharvest quality of strawberry fruit. *Postharvest Biol. Technol.* **2018**, *139*, 47–55. [CrossRef]
46. Xu, L.; Yue, Q.; Bian, F.; Sun, H.; Zhai, H.; Yao, Y. Melatonin Enhances Phenolics Accumulation Partially via Ethylene Signaling and Resulted in High Antioxidant Capacity in Grape Berries. *Front. Plant Sci.* **2017**, *8*, 1426. [CrossRef] [PubMed]
47. Wang, L.; Luo, Z.; Ban, Z.; Jiang, N.; Yang, M.; Li, L. Role of exogenous melatonin involved in phenolic metabolism of Zizyphus jujuba fruit. *Food Chem.* **2021**, *341*, 128268. [CrossRef] [PubMed]
48. Sun, C.; Huang, Y.; Lian, S.; Saleem, M.; Li, B.; Wang, C. Improving the biocontrol efficacy of Meyerozyma guilliermondii Y-1 with melatonin against postharvest gray mold in apple fruit. *Postharvest Biol. Technol.* **2021**, *171*, 111351. [CrossRef]
49. Zhang, S.; Zheng, X.; Reiter, R.J.; Feng, S.; Wang, Y.; Liu, S.; Jin, L.; Li, Z.; Datla, R.; Ren, M. Melatonin Attenuates Potato Late Blight by Disrupting Cell Growth, Stress Tolerance, Fungicide Susceptibility and Homeostasis of Gene Expression in Phytophthora infestans. *Front. Plant Sci.* **2017**, *8*, 1993. [CrossRef] [PubMed]
50. Yin, X.; Wei, Y.; Song, W.; Zhang, H.; Liu, G.; Chen, Y.; Li, L.-Z.; Alolga, R.N.; Ma, G.; Reiter, R.J.; et al. Melatonin as an inducer of arecoline and their coordinated roles in anti-oxidative activity and immune responses. *Food Funct.* **2020**, *11*, 8788–8799. [CrossRef]
51. Liu, C.; Chen, L.; Zhao, R.; Li, R.; Zhang, S.; Yu, W.; Sheng, J.; Shen, L. Melatonin Induces Disease Resistance to Botrytis cinerea in Tomato Fruit by Activating Jasmonic Acid Signaling Pathway. *J. Agric. Food Chem.* **2019**, *67*, 6116–6124. [CrossRef]
52. Lin, Y.; Fan, L.; Xia, X.; Wang, Z.; Yin, Y.; Cheng, Y.; Li, Z. Melatonin decreases resistance to postharvest green mold on citrus fruit by scavenging defense-related reactive oxygen species. *Postharvest Biol. Technol.* **2019**, *153*, 21–30. [CrossRef]
53. Wang, L.; Luo, Z.; Yang, M.; Li, D.; Qi, M.; Xu, Y.; Abdelshafy, A.M.; Ban, Z.; Wang, F.; Li, L. Role of exogenous melatonin in table grapes: First evidence on contribution to the phenolics-oriented response. *Food Chem.* **2020**, *329*, 127155. [CrossRef] [PubMed]
54. Michailidis, M.; Tanou, G.; Sarrou, E.; Karagiannis, E.; Ganopoulos, I.; Martens, S.; Molassiotis, A. Pre- and Post-harvest Melatonin Application Boosted Phenolic Compounds Accumulation and Altered Respiratory Characters in Sweet Cherry Fruit. *Front. Nutr.* **2021**, *8*, 306. [CrossRef] [PubMed]

foods

MDPI

Article

Hydrogen Sulfide Treatment Alleviates Chilling Injury in Cucumber Fruit by Regulating Antioxidant Capacity, Energy Metabolism and Proline Metabolism

Jingda Wang, Yaqin Zhao, Zhiqian Ma, Yonghua Zheng and Peng Jin *

College of Food Science and Technology, Nanjing Agricultural University, Nanjing 210095, China
* Correspondence: pjin@njau.edu.cn; Tel.: +86-25-84395315; Fax: +86-25-84395618

Abstract: Although low-temperature storage could maintain the quality of fruits and vegetables, it may also result in chilling injury (CI) in cold-sensitive produce, such as cucumbers. This can seriously affect their quality." The antioxidant capacity, energy metabolism and proline metabolism of cucumbers treated with hydrogen sulfide (H_2S) were studied in this assay. The outcomes displayed that H_2S treatment effectively reduced CI and delayed the increase in electrolyte leakage (EL) and malondialdehyde (MDA) content. In addition, the H_2S-treated cucumber fruit exhibited higher L* and hue angle values, as well as nutrients such as ascorbic acid (AsA). The H_2S-treated fruit showed lower levels of reactive oxygen species (ROS) and higher antioxidant enzyme activities. Meanwhile, H_2S treatment also increased the activities of the essential enzymes involved in energy metabolism, including cytochrome C oxidase (CCO), succinate dehydrogenase (SDH), H^+-ATPase and Ca^{2+}-ATPase, which improved the energy supply. H_2S induced higher ornithine δ-aminotransferase (OAT) and Δ-1-pyrroline-5-carboxylate synthetase (P5CS) activities, and reduced proline dehydrogenase (PDH) activity, promoting the accumulation of proline. These results indicated that H_2S could alleviate CI in the cucumber fruit by modulating antioxidant capacity, energy metabolism and proline metabolism, thereby extending the shelf life of postharvest cucumbers.

Keywords: cucumber; hydrogen sulfide; chilling injury; antioxidant enzyme; energy; proline

Citation: Wang, J.; Zhao, Y.; Ma, Z.; Zheng, Y.; Jin, P. Hydrogen Sulfide Treatment Alleviates Chilling Injury in Cucumber Fruit by Regulating Antioxidant Capacity, Energy Metabolism and Proline Metabolism. *Foods* **2022**, *11*, 2749. https://doi.org/10.3390/foods11182749

Academic Editor: Eleni Tsantili

Received: 18 July 2022
Accepted: 31 August 2022
Published: 7 September 2022

1. Introduction

Cucumber (*Cucumis sativus* L.) belongs to the Cucurbitaceae family and is a popular fruit that is rich in nutrients. Cucumber fruit is a good source of antioxidants, dietary fiber and ascorbic acid (AsA) [1]. Cold storage is one of the most important techniques applied to prevent degradation, diminish respiration and extend the preservation period of cucumber. However, cucumber is a very temperature-sensitive commodity. When the temperature is lower than 10 °C, many chilling injury (CI) symptoms, such as surface pitting, water spots, scale depressions, tissue collapse and decay, are prone to occur in cucumber [2]. CI symptoms reduce the shelf life of cucumber fruit and caused huge economic losses. For many years, many approaches have been applied to reduce CI symptoms in cucumber fruit, including salicylic acid [3], heat stimulation [4], melatonin [5] and putrescine [6]. Hydrogen sulfide (H_2S) is universally recognized to be a poisonous gas that smells like rotten eggs. Nevertheless, there is growing evidence that only high concentrations of H_2S are harmful to cells, while low levels of endogenous H_2S perform a diverse range of physiological functions in horticultural products [7]. H_2S is considered to be the third clock gas signaling molecule alongside nitric oxide (NO) and carbon monoxide (CO) [8]. Previous findings suggested that H_2S could reduce CI in bananas [9], sweet cherries [10] and hawthorns [11]. Therefore, H_2S is considered to have a potential role in mitigating chilling injury in cucumbers.

The intracellular energy status is thought to be a major factor in the post-harvest maturation and senescence of horticulture products [12,13]. More and more studies have shown that an enhanced energy state sustains membrane integrity [14,15]. The cell membrane is the main site of CI development [16]. Maintaining higher levels of adenosine triphosphate (ATP) and energy charge (EC) also contributes to membrane integrity [17]. Physiological responses associated with senescence, for example, enhanced membrane permeability and increased reactive oxygen species (ROS) generation, are likely to be associated with inadequate energy supply. An increasing amount of research showed that energy metabolism was linked to the occurrence of CI in plants. Under low-temperature stress, plant tissues required more energy to maintain normal life activities. However, the tissues had a decreased supply of energy due to the reduced metabolic level. The contradiction between these two aspects might contribute to CI in horticulture products [13]. Proline is an osmoregulatory substance that is widely present in plants [18]. When plants are under cold stress, proline content increases significantly to protect their tissues [19]. The accumulation of proline has a major function in cold resistance in plants [20].

However, the impacts of exogenous H_2S treatment on antioxidant metabolism, energy metabolism and proline metabolism of cucumber fruit have not been reported. Thus, the aim of this research was to study the effect of H_2S treatment on cucumber fruit and the mechanism associated with mitigating CI.

2. Materials and Methods

2.1. Plant Materials and Treatments

Cucumber fruit were picked at the commercial maturity stage from a farm in Nanjing, Jiangsu, China. Then they were immediately carried back to the laboratory. Cucumbers of uniform size and no mechanical injury were picked and randomly separated into two groups (300 fruit in each group). The control group was soaked in deionized water for 10 min, while the treatment group was soaked in a 1.0 mM sodium hydrosulfide (NaHS) for 10 min. All fruit were packed in plastic bags after drying, with 5 fruit in each bag. Then, they were placed in a thermostat at a temperature of 4 °C and relative humidity of 85–90%. Sixty fruit were randomly selected every 3 days during the storage period with 3 replicates. The cucumber fruit were stored for a total of 15 days. On each sampling day, 30 fruit were placed on a rack at 20 °C for 2 days to determine the CI index. Another 30 fruit were cut into small pieces, frozen using liquid nitrogen and stored at −80 °C for the determination of other indicators.

2.2. CI index, Electrolyte Leakage (EL), L* Value and Hue Angle (H*) Value

The CI index of cucumber was measured based on the method of Lin et al. [21]. The fruit were ranked on the CI scale from 0 to 4, where 0—none, 1—slight, 2—moderate, 3—moderately severe and 4—severe. The CI index is presented in Equation (1):

$$\text{CI Index} = \frac{\sum(\text{CI scale} \times \text{the number of fruit in each scale})}{(4 \times \text{the total number of fruit})} \tag{1}$$

The measurement of EL was based on the method of Ali et al. [22]. Ten small pieces of the middle part of cucumber peels were taken. The thickness of each piece was about 1 mm. The piece was placed in a test tube. The conductivity was determined with a conductivity meter after adding 25 mL of distilled water. Conductivity C_1 was measured after standing for 30 min and then the test tube was boiled for 15 min. Conductivity C_2 was measured after cooling. The formula is as follows (Equation (2)):

$$\text{EL}(\%) = \frac{C_1 - C_0}{C_2 - C_0} \times 100\% \tag{2}$$

The L* and H* values reflect the lightness and yellow-green values of the color, respectively. Three points were picked up on the equator of each cucumber fruit and measured with a color difference meter. The value of H* was calculated using Equation (3):

$$H^* = 180 - \arctan\left(\frac{a^*}{b^*}\right) \quad (3)$$

2.3. Chlorophyll, Ascorbic Acid and Malondialdehyde (MDA) Content

Measurement of the chlorophyll content of cucumber peel was based on the method of Zhang et al. [3]. Two grams of cucumber peel was placed in a mortar, ground well with 80% acetone, left to stand and filtered. The absorbances at the wavelengths of 663 nm and 645 nm were measured using a spectrophotometer. The results were expressed as the content of chlorophyll contained in each gram of the sample, namely, mg g^{-1} FW. The formula for calculating the chlorophyll content was as follows (Equation (4)):

$$\text{Chlorophyll content (mg/g)} = (20.29 \times A_{645} + 8.05 \times A_{663})\ /\ 1000 \quad (4)$$

A_{645} and A_{663} represent the absorbance values at wavelengths of 645 nm and 663 nm, respectively.

The AsA content was assayed according to Jia et al. [6]. Two grams of cucumber pulp was weighed. Then, the pulp was homogenized. The absorbance of the reactive solution at 534 nm was calculated using the o-phenanthroline method. The results were expressed as mg g^{-1} FW.

The MDA content was lightly modified based on the method of Jin et al. [23]. The absorbances at 600 nm, 532 nm and 450 nm were measured. The content of MDA was expressed as nmol g^{-1} FW. The formula for calculating MDA content is as follows (Equation (5)):

$$\text{MDA content}\left(\frac{\text{nmol}}{\text{g}}\right) = 6.45 \times (A_{532} - A_{600}) - 0.56 \times A_{450} \quad (5)$$

A_{532}, A_{600} and A_{450} represent the absorbance values at wavelengths of 645 nm, 663 nm and 450 nm, respectively.

2.4. 1,1-Diphenyl-2-Picrylhydrazyl (DPPH·) Scavenging Rate, Hydroxyl Radical (·OH) Scavenging Rate, Superoxide Anion ($O_2^{\cdot-}$) Production Rate and Hydrogen Peroxide (H_2O_2) Content

The DPPH· scavenging rate was measured according to Al Ubeed et al. [24]. The formula was as follows (Equation (6)):

$$\text{DPPH· scavenging rate (\%)} = \left[\frac{A_0 - A_1}{A_0}\right] \times 100 \quad (6)$$

A_0 and A_1 represent the absorbance of the control and sample, respectively.

The ·OH scavenging rate was measured according to Ma et al. [25]. The results were calculated using the following Equation (7):

$$\text{·OH scavenging rate (\%)} = \left[\frac{A_0 - A_1}{A_0}\right] \times 100 \quad (7)$$

The meanings of A_0 and A_1 are the same as in Equation (5).

$O_2^{\cdot-}$ production was calculated according to Ma et al. [25]. Two grams of pulp tissue was ground in 5 mL of 50 mM phosphate buffer. The homogenate was centrifuged at 10,000× *g* for 20 min at 4 °C. The supernatant was used for the determination of $O_2^{\cdot-}$ production. The absorbance value of the reaction solution at the wavelength of 530 nm was measured. A potassium nitrite (KNO_2) standard curve was used to measure the $O_2^{\cdot -}$ production. The results were expressed in μmol min^{-1} g^{-1} FW.

To measure the content of H_2O_2, 2.0 g of cucumber pulp was mixed with 100% acetone. The H_2O_2 content in the crude extract was detected according to Zhang et al. [26]. The outcome of H_2O_2 content was indicated in μmol g^{-1} FW.

2.5. Ascorbate Peroxidase (APX), Superoxide Dismutase (SOD), Catalase (CAT) and Peroxidase (POD) Activities

The activities of APX, SOD, CAT and POD were determined according to Wang et al. [27]. The amount of enzyme needed to decrease the absorbance at 290 nm by 0.01 per minute was one APX unit (U). The quantity needed to suppress NBT photoreduction by 50% per minute was used as one unit of enzyme activity (U) of SOD. The amount of enzyme that decreased the absorbance value of the reaction system at 240 nm by 0.01 per minute was considered as one CAT unit (U). One POD unit (U) was considered the quantity of enzyme needed to decrease the absorbance value of the reaction solution by 0.01 per minute at 470 nm. All anti-oxidant enzyme activities were expressed in U g^{-1} FW.

2.6. H^+-ATPase, Ca^{2+}-ATPase, Cytochrome C Oxidase (CCO) and Succinate Dehydrogenase (SDH) Activities

The extraction of cucumber mitochondria followed the approach of Li et al. [28]. Cucumber samples (5.0 g each) were ground with 10 mL of pre-chilled Tris-HCl and filtered through five layers of gauze. The supernatant was centrifuged to obtain the precipitate. The precipitate obtained after undergoing centrifugation again for 10 min was the mitochondria. It was placed at a low temperature for the next step of the experiment. The activities of H^+-ATPase and Ca^{2+}-ATPase were determined according to Jin et al. [29]. The measurements of SDH and CCO activities were measured according to Cheng et al. [30]. The results were expressed as U g^{-1} FW.

2.7. ATP, Adenosine Diphosphate (ADP) and Adenosine Monophosphate (AMP) Contents and EC

The energy level was determined according to Zhou et al. [31]. The cucumber pulp was mixed with 5 mL $HClO_4$ solution and thoroughly ground. Then, the volume was increased to 6 mL with ultra-pure water. It was filtered with a 0.45 μm water filter and analyzed using high-performance liquid chromatography. The samples were qualitatively analyzed according to the retention time of the standard and quantitatively analyzed according to the peak area of the standard. The final content was expressed in μg g^{-1} FW. The calculation method for EC is as follows (Equation (8)):

$$EC(\%) = \frac{(ATP + 0.5 \times ADP)}{(ATP + ADP + AMP)} \times 100 \tag{8}$$

2.8. Proline Content and Δ-1-Pyrroline-5-Carboxylate Synthetase (P5CS), Proline Dehydrogenase (PDH) and Ornithine δ-Aminotransferase (OAT) Activities

Proline content was assayed according to Zuo et al. [18]. Cucumber samples (1.0 g each) were ground with 3% sulfosalicylic acid. Samples were extracted by shaking at 100 °C for 10 min after grinding into a homogenate. After centrifugation, the supernatant was extracted by adding acidic ninhydrin and boiled for 25 min. The reaction solution was extracted with toluene. The absorbance value of the organic phase at 520 nm was measured. The results were expressed in μg g^{-1} FW. The enzymatic activities of P5CS, PDH and OAT were measured according to Zhang et al. [26]. The enzyme solution was mixed with the reaction system via centrifugation at 12,000× *g*. The absorbance data at 340 nm were measured. The change in absorbance value by 0.001 per minute of the mixed reaction solution was considered as 1 U of enzyme activity. The results were shown as U g^{-1} FW. Reactions were initiated with 0.15 mL NAD^+ for the determination of PDH activity. Cucumber samples were triturated with sodium phosphate buffer to measure the OAT activity. The supernatant obtained via centrifugation was mixed with the reaction system. The absorbance values at 510 nm were measured.

2.9. Statistical Analysis

Experiments were performed in a completely randomized design and the results were expressed as mean ± standard deviation (SD). Origin 2022 software was used for graphing, and SAS 9.21 software was used for independent samples *t*-tests ($p < 0.05$).

3. Results

3.1. CI Index, EL, L* Value and H* Value

The CI index increased during storage. The symptoms of CI in the control group appeared earlier and were more severe than in the H_2S-treated group (Figure 1A). EL from cucumber fruit remained elevated throughout the storage period; the increase in the H_2S treatment group was significantly smaller than in the control group (Figure 1B).

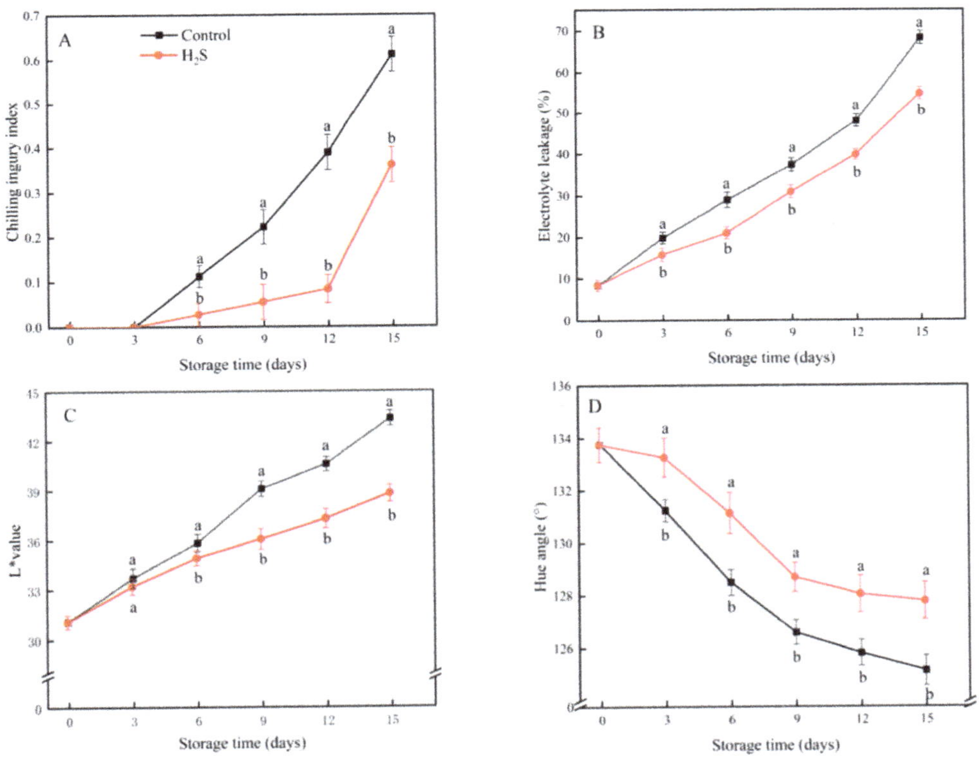

Figure 1. Effects of the H_2S treatment on the chilling injury index (**A**), electrolyte leakage (**B**), L* value (**C**) and hue angle (**D**) of the cucumber stored at 4 °C over the whole storage period. Different letters represent significant differences between treatments on the same sampling day ($p < 0.05$).

The L* value of the cucumber peel showed an increasing trend with extended storage (Figure 1C). The L* value of the H_2S-treated group remained lower than that of the control group except for the third day of storage. The H* value of the cucumber peel showed a decreasing trend with extended storage (Figure 1D). The H* value of the H_2S treatment group remained higher than that of the control group for the whole storage. The L* and H* values of the H_2S treatment group were 89.5% and 102.8% of those of the control group on day 15, respectively.

3.2. Chlorophyll, AsA and MDA Content

The content of chlorophyll gradually declined during storage period and the H_2S treatment significantly suppressed the decrease in chlorophyll content (Figure 2A). The chlorophyll content in the H_2S treatment group was 13.8% higher than in the untreated group on day 15. As shown in Figure 2B, the AsA content in the cucumber fruit in the control group remained stable for the first 6 days of refrigeration and then decreased progressively. Except for the third and sixth days of storage, the contents of AsA in the H_2S group remained higher than that of the control group.

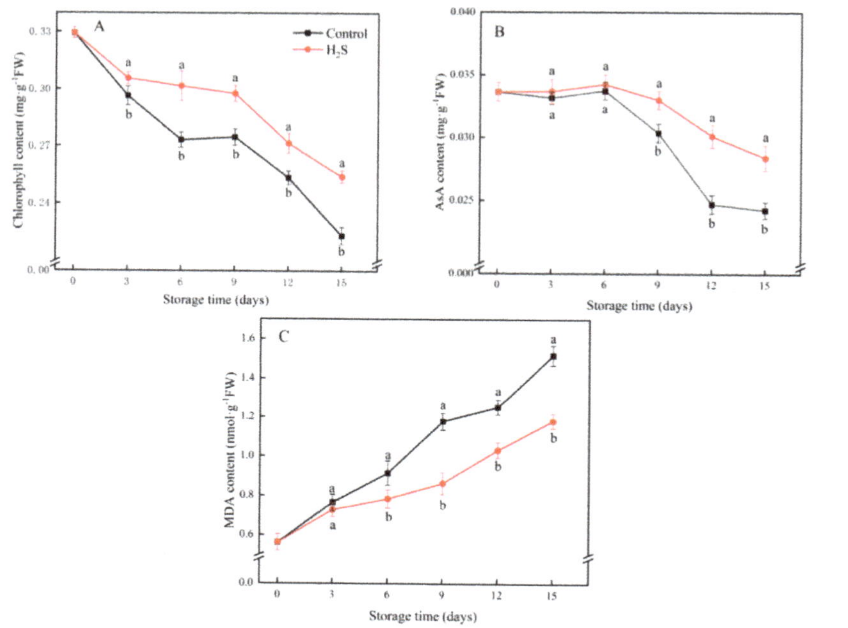

Figure 2. Effects of the H_2S treatment on chlorophyll content (**A**), AsA content (**B**) and MDA content (**C**) of the cucumber fruit stored at 4 °C over the whole storage period. Different letters represent significant differences between treatments on the same sampling day ($p < 0.05$).

The MDA content of the cucumber increased throughout storage. The H_2S treatment significantly suppressed the accumulation of MDA content (Figure 2C). The difference in MDA content in the cucumber fruit between the two groups increased from day 6. MDA content in the control group was 28.9% higher than that in the H_2S treatment group on day 15 of storage.

3.3. DPPH· Scavenging Rate, ·OH Scavenging Rate, $O_2^{\cdot -}$ Production Rate and H_2O_2 Content

The DPPH· scavenging rate in the cucumber fruit initially increased sharply. Its peak was reached on day 6 and declined gently thereafter (Figure 3A). The ·OH radical scavenging rate was elevated continuously before day 9 and then declined gradually (Figure 3B). The DPPH· and ·OH scavenging rates in the cucumber fruit treated with H_2S remained higher than those in the control group.

The $O_2^{\cdot -}$ production rate in the cucumber increased steadily throughout storage. The H_2S treatment significantly slowed the increase in the $O_2^{\cdot -}$ production rate during cold storage (Figure 3C). The H_2O_2 content of the cucumber fruit increased rapidly until day 6 and then began to decrease. Compared with the control group, the H_2O_2 content of the H_2S-treated fruit was consistently lower (Figure 3D). The H_2O_2 content of the H_2S-treated fruit was only 72.9% of the content in the untreated fruit on day 6.

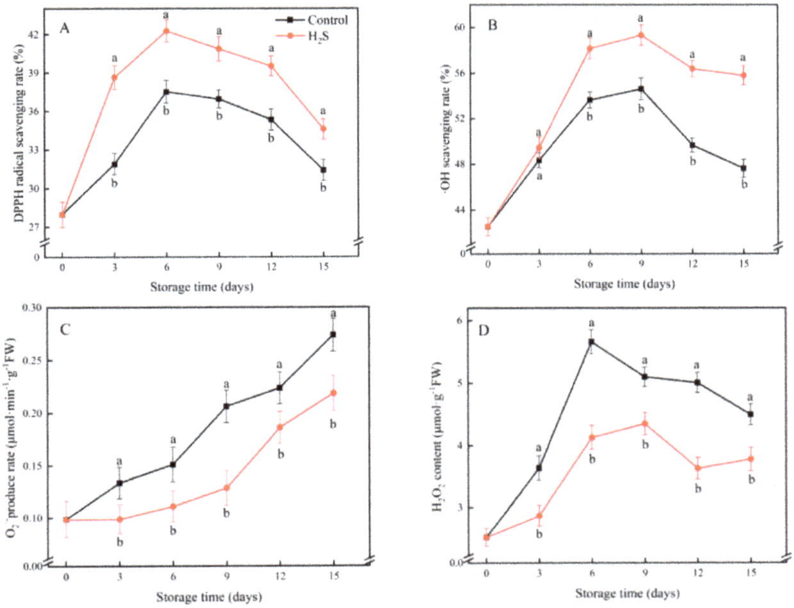

Figure 3. Effects of the H_2S treatment on the DPPH· radical scavenging rate (**A**), ·OH scavenging rate (**B**), $O_2^{\cdot-}$ production rate (**C**) and H_2O_2 content (**D**) of the cucumber fruit stored at 4 °C over the whole storage period. Different letters represent significant differences between treatments on the same sampling day ($p < 0.05$).

3.4. APX, SOD, CAT and POD Activities

In Figure 4, the activities of APX, SOD, CAT and POD displayed an increasing trend initially, but a decrease as storage continued. APX activity peaked on day 9 and then declined rapidly. The control group was always lower than the H_2S-treated group throughout the storage period. The APX activity of the H_2S-treated group was 27.1% higher than the untreated group on day 9 of the cold storage (Figure 4A). The SOD activity of the untreated group was consistently lower than that of the H_2S-treated group during storage. The SOD activity of the H_2S-treated group was 34.8% higher than that of the untreated group on day 15 of cold storage. The H_2S treatment significantly enhanced the CAT activity and suppressed its decrease during the storage period. The CAT activity of the H_2S-treated group was 36.0% higher than that of the control group on day 9 (Figure 4C). The POD activity of the control group was always lower than that of the treatment group (Figure 4D). The difference between the two groups reached a maximum on day 9 of storage. At this time, the activity of the H_2S-treated group was 23.4% higher than that of the control.

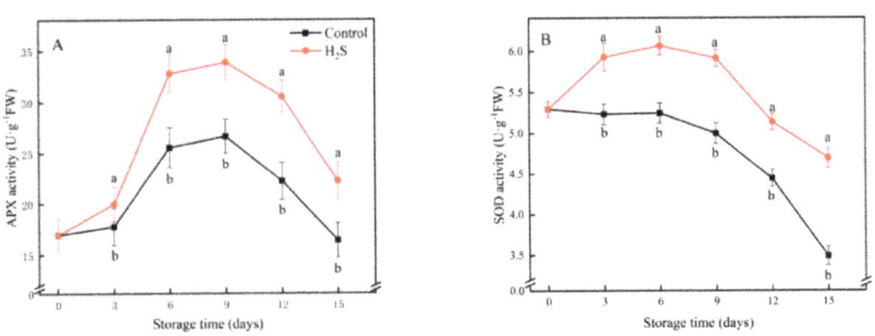

Figure 4. *Cont.*

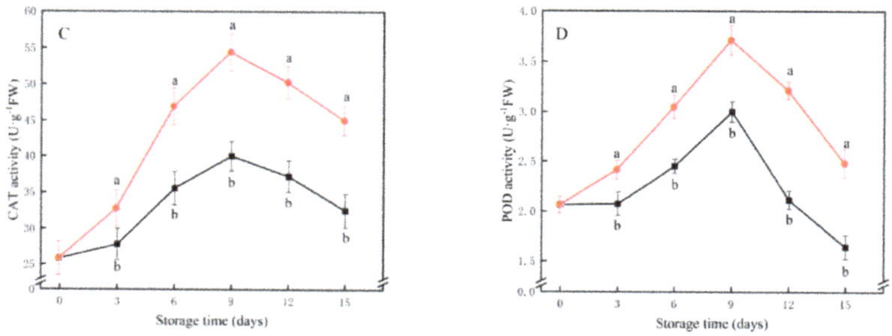

Figure 4. Effects of the H_2S treatment on the activities of APX (**A**), SOD (**B**), CAT (**C**) and POD (**D**) of the cucumber stored at 4 °C over the whole storage period. Different letters represent significant differences between treatments on the same sampling day ($p < 0.05$).

3.5. H^+-ATPase, Ca^{2+}-ATPase, CCO and SDH Activities

The activity of H^+-ATPase increased initially in the cucumber fruit. After reaching its peak, it began to decline (Figure 5A). The H^+-ATPase activity of the H_2S-treated group was significantly higher than that of the untreated fruit. The activity of Ca^{2+}-ATPase initially increased and then started to decrease after day 6 (Figure 5B). The activity of the H_2S treatment group was always maintained at a high level. An analogous tendency was observed in SDH and CCO. The difference was that the activities of SDH and CCO began to decline after day 9 of refrigeration. The H_2S treatment significantly elevated the activities of all the above enzymes. The activities of H^+-ATPase, Ca^{2+}-ATPase, SDH and CCO in the H_2S-treated group were 18.6%, 15.4%, 14.7% and 23.6% higher than those in the untreated group on day 15, respectively.

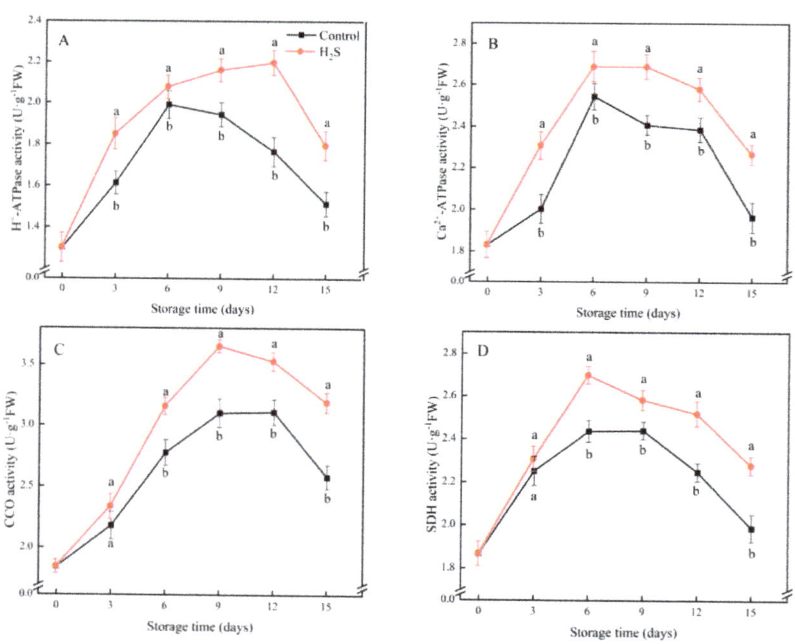

Figure 5. Effects of the H_2S treatment on the activities of H^+-ATPase (**A**), Ca^{2+}-ATPase (**B**), CCO (**C**) and SDH (**D**) of the cucumber stored at 4 °C over the whole storage period. Different letters represent significant differences between treatments on the same sampling day ($p < 0.05$).

3.6. *ATP, ADP and AMP Contents and EC*

The ATP and ADP contents of the cucumber fruit gradually decreased during cold storage (Figure 6A,B). The H_2S treatment significantly maintained the ATP and ADP contents; the H_2S-treated group had 12.6% and 14.6% higher contents than the untreated group on day 15, respectively. The AMP content increased continuously during refrigeration and H_2S treatment could inhibit the increase in AMP in the cucumber fruit (Figure 6C). The AMP content of the H_2S-treated group was 11.8% lower than the untreated group on day 15. According to the changes in ATP, ADP and AMP contents in the cucumber fruit, the energy charge gradually declined over the whole storage period (Figure 6C). The H_2S treatment significantly inhibited the decrease in the EC level.

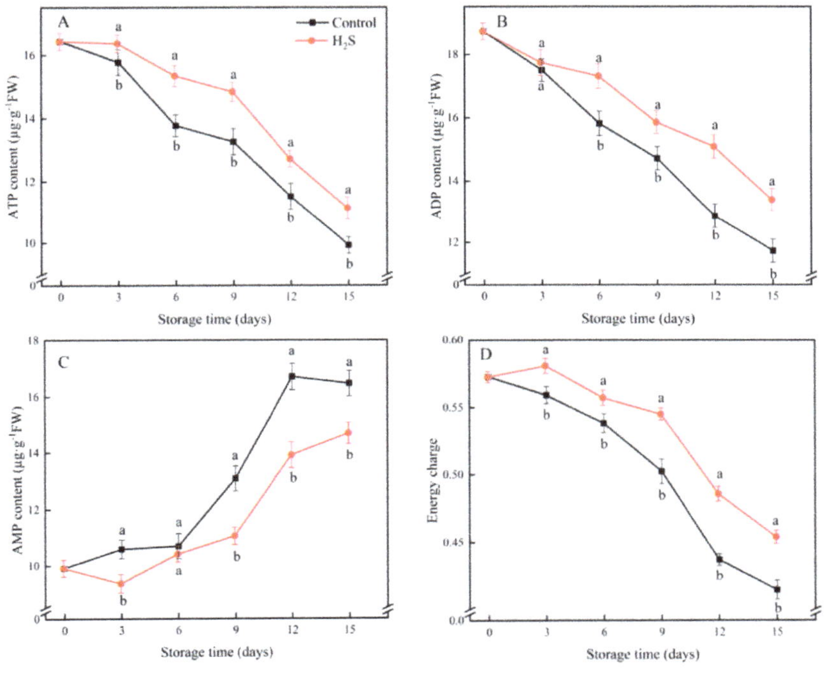

Figure 6. Effects of the H_2S treatment on the activities of ATP (**A**), ADP (**B**), AMP (**C**) and energy charge (**D**) of the cucumber stored at 4 °C over the whole storage period. Different letters represent significant differences between treatments on the same sampling day ($p < 0.05$).

3.7. *Proline Content and P5CS, PDH and OAT Activities*

The proline content increased remarkably during the first 12 days of refrigeration and remained at a high level thereafter (Figure 7A). The H_2S treatment significantly promoted the accumulation of proline. During the low-temperature storage, the P5CS activity gradually increased and was higher in the H_2S-treated group than in the untreated group (Figure 7B). The activity of PDH decreased quickly in the first 9 days of refrigeration and then decreased gently (Figure 7D). The H_2S-treated group showed lower PDH activity compared with the untreated group. The OAT activity increased substantially until day 12, after which it started to decrease (Figure 7C). The H_2S treatment significantly enhanced the OAT activity. The OAT activity in the treatment group was 16.2% higher than the control group on day 15 of refrigeration.

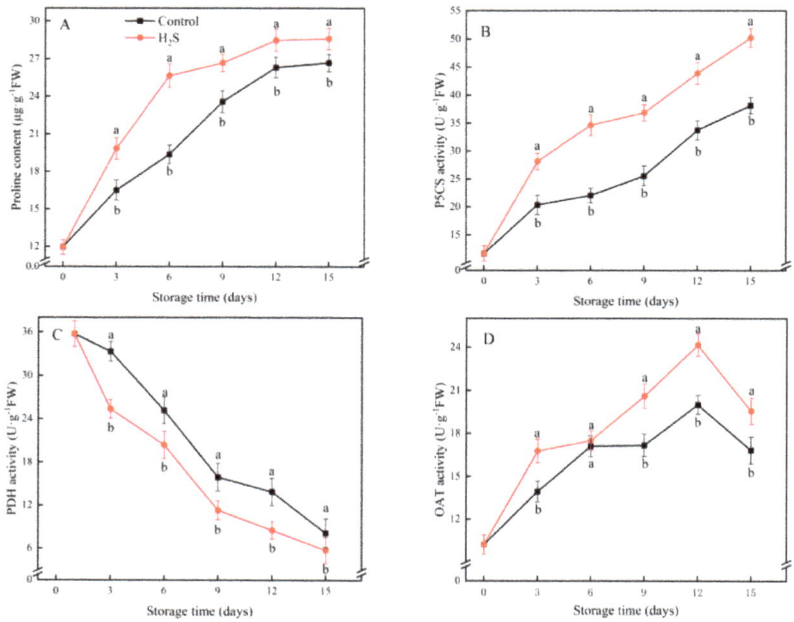

Figure 7. Effects of the H_2S treatment on the proline content (**A**), P5CS activity (**B**), PDH activity (**C**) and OAT activity (**D**) of the cucumber stored at 4 °C over the whole storage period. Different letters represent significant differences between treatments on the same sampling day ($p < 0.05$).

4. Discussion

In plants, the production of ROS and its detoxification are in homeostasis, thus maintaining low levels of ROS [32]. However, during the senescence of fruits and vegetables or when they are influenced by biotic or abiotic stresses, the homeostatic system of ROS generation and scavenging is disrupted [25]. This, in turn, leads to membrane lipid peroxidation and the disruption of biofilm structure and function. Eventually, CI symptoms are induced. Plants have enzymatic and non-enzymatic antioxidant systems that eliminate ROS [9]. The enzymatic antioxidant system in plants involves SOD, POD, APX and CAT, which effectively scavenge free radicals, such as $O_2^{\cdot -}$ and H_2O_2, in plants [33]. SOD participates in the first level of resistance against ROS by scavenging $O_2^{\cdot -}$ from cells to generate H_2O_2 and O_2 [4]. H_2O_2 is then catalyzed by CAT and POD to generate H_2O and O_2 [11]. APX plays a catalytic function in the ascorbate–glutathione (AsA–GSH) cycle and their combined action can effectively scavenge ROS [33]. The non-enzymatic antioxidant system of plants includes various compounds, such as AsA GSH, phenolics, carotenoids and others [34]. Therefore, they can prevent the poisonous effects of ROS on cells. Many reports revealed that the anti-oxidant ability of fruits and vegetables was related to cold resistance. For instance, Aghdam et al. [11] discovered that H_2S treatment could effectively increase the activities of SOD, CAT and APX in hawthorn while reducing the symptoms of hawthorn CI. Chen et al. [12] noted that 6-Benzylaminopurine treatment improved SOD, CAT and APX activities in cucumber fruit while reducing CI symptoms in cucumber. Luo et al. [9] reported that H_2S treatment diminished oxidative damage in cucumber via increasing the activity of antioxidant enzymes to reduce ROS. In this study, the activities of antioxidant enzymes, such as SOD, CAT, APX and POD, and the scavenging rates of DPPH· and ·OH in the cucumber fruit of the H_2S-treated group remained higher than those in the untreated group, while the contents of $O_2^{\cdot -}$ and H_2O_2 were significantly lower than those in the untreated group. This indicated that H_2S could effectively improve the antioxidant capacity of cucumber fruit, thereby reducing chilling injury.

The intracellular energy supply is an essential factor in the control of fruit ripening and senescence [23]. Reduced energy levels can have negative effects on various fruits, such as physiological disorders [17]. Previous studies revealed that insufficient energy supply is related to the occurrence of CI in plants [28]. For example, the ATP content and EC level in banana declined continuously during refrigeration and the storage quality was also decreased. However, after H_2S treatment, its ATP content and EC were improved and the symptoms of CI were relieved [35]. Cheng et al. [30] found that 1-methylcyclopropene (1-MCP) could enhance cold resistance in pear by improving enzyme activities linked to energy metabolism and keeping higher levels of ATP and EC. In this experiment, the H_2S-treated group maintained higher ATP, ADP and EC levels and lower AMP content compared with the control group. This showed that H_2S treatment could improve the cold resistance of cucumber by keeping a higher level of intracellular EC.

To better illustrate the involvement of H_2S in the regulation of energy metabolism, the activities of four enzymes engaged in ATP synthesis were investigated. H^+-ATPase, Ca^{2+}-ATPase, SDH and CCO are essential enzymes of energy metabolism [31]. The trends in the activities of these enzymes reflect the function of mitochondria and their ability to synthesize energy [23]. Among them, ATPase is a type of enzyme that releases more energy by catalyzing the decomposition of ATP into ADP and phosphate ions [30]. H^+-ATPase provides energy and power for the transport of various nutrients and ions across the membrane and maintains a relatively stable cytoplasmic pH [29]. The decrease in H^+-ATPase activity is positively correlated with cellular energy loss. Ca^{2+}-ATPase can take energy released from ATP to transfer Ca^{2+} from the cytoplasm to the mitochondria or vesicles, thus maintaining the dynamic balance of Ca^{2+} concentration and contributing to the maintenance of cell homeostasis [28]. Shifts in SDH and CCO activity have an impact on the flow of electron transport chains, which, in turn, affects energy metabolism [13]. The deactivation of energy metabolizing enzymes leads to mitochondrial dysfunction, inadequate energy production and cell death [31]. Hence, energy deficiency is related to plant CI. H_2S contributes to the cold tolerance of banana fruit during storage and is associated with increased H^+-ATPase, Ca^{2+}-ATPase, SDH and CCO activities [35]. Liu et al. [19] reported that brassinolide could alleviate CI in bamboo shoots by enhancing enzyme activities related to energy metabolism and retaining higher levels of ATP and EC. In the current research, the activities of four key enzymes of energy metabolism were significantly higher in the H_2S-treated group than in the untreated group, which maintained higher ATP content and EC level of the cucumber. Thus, the tolerance of the cucumber fruit to low temperature during storage was improved and the occurrence of CI of the cucumber fruit was further inhibited.

Proline is an osmoregulatory substance with hydration capacity [19]. It not only maintains the osmotic balance of cells under adversity but also protects various metabolic enzymes in cells and regulates the physiological activities of plants [9]. The main proline production pathways in plant cells are P5CS and OAT, which catalyze the synthesis of proline from glutamate and ornithine, respectively. Its main depletion pathway is PDH, which can catalyze proline degradation [20]. Many studies demonstrated that many fruit and vegetable preservation methods could increase the proline content during postharvest storage by regulating the activity of these three enzymes. It stabilizes the cellular osmotic pressure balance, which, in turn, improves the cold resistance of horticulture products. Regulation of the activity of these three enzymes induced proline accumulation, which then reduced CI in banana [9], loquat [20] and bamboo shoots [19]. Similarly, in our study, the H_2S-treated cucumbers showed higher OAT and P5CS activities and reduced PDH activity. This enhanced the accumulation of proline, and thus, reduced the presence of CI.

5. Conclusions

The H_2S treatment improved the activities of antioxidant enzymes and maintained the ROS at a low level, thereby reducing the amount of oxidative damage to cell membranes. At the same time, the H_2S treatment could effectively improve the activity of key enzymes of energy metabolism, inhibit the decrease of ATP and ADP content and the increase of

AMP content, and maintain a high EC level. The H_2S treatment also promoted proline accumulation by regulating proline metabolism. In conclusion, H_2S enhanced the tolerance of the cucumber fruit to CI, which may be related to the regulation of antioxidants, energy and proline metabolism.

Author Contributions: Conceptualization, J.W. and P.J.; data curation, J.W., Z.M. and Y.Z. (Yaqin Zhao); funding acquisition, P.J.; investigation, J.W. and Y.Z. (Yaqin Zhao); methodology, J.W. and Y.Z. (Yonghua Zheng); supervision, P.J.; validation, P.J. and Y.Z. (Yonghua Zheng); writing—original draft, J.W. All authors have read and agreed to the published version of the manuscript.

Funding: This study was supported by the National Key R&D Program of China (2021YFD2100502-4).

Data Availability Statement: Data is contained within the article.

Conflicts of Interest: The authors declare no conflict of interest.

References

1. Kahramanoğlu, İ.; Usanmaz, S. Improving postharvest storage quality of cucumber fruit by modified atmosphere packaging and biomaterials. *HortScience* **2019**, *54*, 2005–2014. [CrossRef]
2. Murad, H.; Nyc, M.A. Evaluating the potential benefits of cucumbers for improved health and skin care. *J. Aging Res. Clin. Pract.* **2016**, *5*, 139–141. [CrossRef]
3. Zhang, Y.; Zhang, M.; Yang, H. Postharvest chitosan-g-salicylic acid application alleviates chilling injury and preserves cucumber fruit quality during cold storage. *Food Chem.* **2015**, *174*, 558–563. [CrossRef] [PubMed]
4. Nasef, I.N. Short hot water as safe treatment induces chilling tolerance and antioxidant enzymes, prevents decay and maintains quality of cold-stored cucumbers. *Postharvest Biol. Technol.* **2018**, *138*, 1–10. [CrossRef]
5. Madebo, M.P.; Luo, S.-m.; Wang, L.; Zheng, Y.-h.; Jin, P. Melatonin treatment induces chilling tolerance by regulating the contents of polyamine, γ-aminobutyric acid, and proline in cucumber fruit. *J. Integr. Agric.* **2021**, *20*, 3060–3074. [CrossRef]
6. Jia, B.; Zheng, Q.; Zuo, J.; Gao, L.; Wang, Q.; Guan, W.; Shi, J. Application of postharvest putrescine treatment to maintain the quality and increase the activity of antioxidative enzyme of cucumber. *Sci. Hortic.* **2018**, *239*, 210–215. [CrossRef]
7. Hancock, J.T.; Whiteman, M. Hydrogen sulfide and cell signaling: Team player or referee? *Plant Physiol. Biochem.* **2014**, *78*, 37–42. [CrossRef] [PubMed]
8. Wang, R. Physiological implications of hydrogen sulfide: A whiff exploration that blossomed. *Physiol. Rev.* **2012**, *92*, 791–896. [CrossRef] [PubMed]
9. Luo, Z.; Li, D.; Du, R.; Mou, W. Hydrogen sulfide alleviates chilling injury of banana fruit by enhanced antioxidant system and proline content. *Sci. Hortic.* **2015**, *183*, 144–151. [CrossRef]
10. Zhi, H.; Dong, Y. Effect of hydrogen sulfide on surface pitting and related cell wall metabolism in sweet cherry during cold storage. *J. Appl. Bot. Food Qual.* **2018**, *91*, 109–113.
11. Aghdam, M.S.; Mahmoudi, R.; Razavi, F.; Rabiei, V.; Soleimani, A. Hydrogen sulfide treatment confers chilling tolerance in hawthorn fruit during cold storage by triggering endogenous H_2S accumulation, enhancing antioxidant enzymes activity and promoting phenols accumulation. *Sci. Hortic.* **2018**, *238*, 264–271. [CrossRef]
12. Chen, B.; Yang, H. 6-Benzylaminopurine alleviates chilling injury of postharvest cucumber fruit through modulating antioxidant system and energy status. *J. Sci. Food Agric.* **2013**, *93*, 1915–1921. [CrossRef] [PubMed]
13. Wang, H.; Qian, Z.; Ma, S.; Zhou, Y.; Patrick, J.W.; Duan, X.; Jiang, Y.; Qu, H. Energy status of ripening and postharvest senescent fruit of litchi (*Litchi chinensis* Sonn.). *BMC Plant Biol.* **2013**, *13*, 55. [CrossRef]
14. Yi, C.; Jiang, Y.; Shi, J.; Qu, H.; Xue, S.; Duan, X.; Shi, J.; Prasad, N.K. ATP-regulation of antioxidant properties and phenolics in litchi fruit during browning and pathogen infection process. *Food Chem.* **2010**, *118*, 42–47. [CrossRef]
15. Shan, T.; Jin, P.; Zhang, Y.; Huang, Y.; Wang, X.; Zheng, Y. Exogenous glycine betaine treatment enhances chilling tolerance of peach fruit during cold storage. *Postharvest Biol. Technol.* **2016**, *114*, 104–110. [CrossRef]
16. Ma, M.; Zhu, Z.; Cheng, S.; Zhou, Q.; Zhou, X.; Kong, X.; Hu, M.; Yin, X.; Wei, B.; Ji, S. Methyl jasmonate alleviates chilling injury by regulating membrane lipid composition in green bell pepper. *Sci. Hortic.* **2020**, *266*, 109308. [CrossRef]
17. Cao, S.; Cai, Y.; Yang, Z.; Joyce, D.C.; Zheng, Y. Effect of MeJA treatment on polyamine, energy status and anthracnose rot of loquat fruit. *Food Chem.* **2014**, *145*, 86–89. [CrossRef]
18. Zuo, X.; Cao, S.; Zhang, M.; Cheng, Z.; Cao, T.; Jin, P.; Zheng, Y. High relative humidity (HRH) storage alleviates chilling injury of zucchini fruit by promoting the accumulation of proline and ABA. *Postharvest Biol. Technol.* **2021**, *171*, 111344. [CrossRef]
19. Liu, Z.; Li, L.; Luo, Z.; Zeng, F.; Jiang, L.; Tang, K. Effect of brassinolide on energy status and proline metabolism in postharvest bamboo shoot during chilling stress. *Postharvest Biol. Technol.* **2016**, *111*, 240–246. [CrossRef]
20. Szabados, L.; Savouré, A. Proline: A multifunctional amino acid. *Trends Plant Sci.* **2010**, *15*, 89–97. [CrossRef]
21. Lin, X.; Wang, L.; Hou, Y.; Zheng, Y.; Jin, P. A combination of melatonin and ethanol treatment improves postharvest quality in bitter melon fruit. *Foods* **2020**, *9*, 1376. [CrossRef]

22. Ali, S.; Khan, A.S.; Malik, A.U. Postharvest L-cysteine application delayed pericarp browning, suppressed lipid peroxidation and maintained antioxidative activities of litchi fruit. *Postharvest Biol. Technol.* **2016**, *121*, 135–142. [CrossRef]
23. Jin, P.; Zhu, H.; Wang, L.; Shan, T.; Zheng, Y. Oxalic acid alleviates chilling injury in peach fruit by regulating energy metabolism and fatty acid contents. *Food Chem.* **2014**, *161*, 87–93. [CrossRef]
24. Al Ubeed, H.M.S.; Wills, R.B.H.; Bowyer, M.C.; Golding, J.B. Interaction of the hydrogen sulphide inhibitor, propargylglycine (PAG), with hydrogen sulphide on postharvest changes of the green leafy vegetable, pak choy. *Postharvest Biol. Technol.* **2019**, *147*, 54–58. [CrossRef]
25. Ma, Y.; Hu, S.; Chen, G.; Zheng, Y.; Jin, P. Cold shock treatment alleviates chilling injury in peach fruit by regulating antioxidant capacity and membrane lipid metabolism. *Food Qual. Saf.* **2022**, *6*, fyab026. [CrossRef]
26. Zhang, Y.; Jin, P.; Huang, Y.; Shan, T.; Wang, L.; Li, Y.; Zheng, Y. Effect of hot water combined with glycine betaine alleviates chilling injury in cold-stored loquat fruit. *Postharvest Biol. Technol.* **2016**, *118*, 141–147. [CrossRef]
27. Wang, L.; Bokhary, S.U.F.; Xie, B.; Hu, S.; Jin, P.; Zheng, Y. Biochemical and molecular effects of glycine betaine treatment on membrane fatty acid metabolism in cold stored peaches. *Postharvest Biol. Technol.* **2019**, *154*, 58–69. [CrossRef]
28. Li, D.; Li, L.; Ge, Z.; Limwachiranon, J.; Ban, Z.; Yang, D.; Luo, Z. Effects of hydrogen sulfide on yellowing and energy metabolism in broccoli. *Postharvest Biol. Technol.* **2017**, *129*, 136–142. [CrossRef]
29. Jin, P.; Zhu, H.; Wang, J.; Chen, J.; Wang, X.; Zheng, Y. Effect of methyl jasmonate on energy metabolism in peach fruit during chilling stress. *J. Sci. Food Agric.* **2013**, *93*, 1827–1832. [CrossRef] [PubMed]
30. Cheng, S.; Wei, B.; Zhou, Q.; Tan, D.; Ji, S. 1-Methylcyclopropene alleviates chilling injury by regulating energy metabolism and fatty acid content in 'Nanguo'pears. *Postharvest Biol. Technol.* **2015**, *109*, 130–136. [CrossRef]
31. Zhou, Q.; Zhang, C.; Cheng, S.; Wei, B.; Liu, X.; Ji, S. Changes in energy metabolism accompanying pitting in blueberries stored at low temperature. *Food Chem.* **2014**, *164*, 493–501. [CrossRef]
32. Liu, H.; Jiang, W.; Cao, J.; Ma, L. A combination of 1-methylcyclopropene treatment and intermittent warming alleviates chilling injury and affects phenolics and antioxidant activity of peach fruit during storage. *Sci. Hortic.* **2018**, *229*, 175–181. [CrossRef]
33. Hou, Y.; Li, Z.; Zheng, Y.; Jin, P. Effects of $CaCl_2$ treatment alleviates chilling injury of loquat fruit (*Eribotrya japonica*) by modulating ROS homeostasis. *Foods* **2021**, *10*, 1662. [CrossRef]
34. Mittler, R. ROS are good. *Trends Plant Sci.* **2017**, *22*, 11–19. [CrossRef]
35. Li, D.; Limwachiranon, J.; Li, L.; Du, R.; Luo, Z. Involvement of energy metabolism to chilling tolerance induced by hydrogen sulfide in cold-stored banana fruit. *Food Chem.* **2016**, *208*, 272–278. [CrossRef] [PubMed]

foods

MDPI

Article

Maintenance of Postharvest Quality and Reactive Oxygen Species Homeostasis of Pitaya Fruit by Essential Oil *p*-Anisaldehyde Treatment

Yanmei Xu [1], Zhijun Cai [2,*], Liangjie Ba [3], Yonghua Qin [1], Xinguo Su [4], Donglan Luo [3], Wei Shan [1], Jianfei Kuang [1], Wangjin Lu [1], Liling Li [1], Jianye Chen [1] and Yating Zhao [1,*]

1 State Key Laboratory for Conservation and Utilization of Subtropical Agro-Bioresources, Guangdong Provincial Key Laboratory of Postharvest Science of Fruits and Vegetables, Engineering Research Center of Southern Horticultural Products Preservation, Ministry of Education, South China Agricultural University, Guangzhou 510642, China; yljddcr123@163.com (Y.X.); qinyh@scau.edu.cn (Y.Q.); shanwei@scau.edu.cn (W.S.); jfkuang@scau.edu.cn (J.K.); wjlu@scau.edu.cn (W.L.); Meilingen@163.com (L.L.); chenjianye@scau.edu.cn (J.C.)
2 College of Food and Drug, Liaoning Agricultural Technical College, Yingkou 115009, China
3 School of Food and Pharmaceutical Engineering, Guizhou Engineering Research Center for Fruit Processing, Guiyang University, Guiyang 550003, China; baliangjie@163.com (L.B.); luodonglan1991@163.com (D.L.)
4 Guangdong AIB Polytechnic, Guangzhou 510507, China; suxg@gdaib.edu.cn
* Correspondence: zhijuncai@126.com (Z.C.); zhaoyating@scau.edu.cn (Y.Z.)

Abstract: The performance of *p*-Anisaldehyde (PAA) for preserving pitaya fruit quality and the underpinning regulatory mechanism were investigated in this study. Results showed that PAA treatment significantly reduced fruit decay, weight loss and loss of firmness, and maintained higher content of total soluble solids, betacyanins, betaxanthins, total phenolics and flavonoids in postharvest pitaya fruits. Compared with control, the increase in hydrogen peroxide (H_2O_2) content and superoxide anion ($O_2^{\bullet-}$) production was inhibited in fruit treated with PAA. Meanwhile, PAA significantly improved the activity of antioxidant enzymes superoxide dismutase (SOD), peroxidase (POD) and catalase (CAT). Moreover, PAA-treated pitaya fruit maintained higher ascorbic acid (AsA) and reduced-glutathione (GSH) content but lower dehydroascorbate (DHA) and oxidized glutathione (GSSG) content, thus sustaining higher ratio of AsA/DHA and GSH/GSSG. In addition, activities of ascorbate peroxidase (APX), glutathione reductase (GR), monodehydroascorbate reductase (MDHAR) and dehydrogenation ascorbic acid reductase (DHAR), as well as the expression of *HpSOD*, *HpPOD*, *HpCAT*, *HpAPX*, *HpGR*, *HpDHAR* and *HpMDHAR*, were enhanced after PAA treatment. The findings suggest that postharvest application of PAA may be a reliable method to control postharvest decay and preserve quality of harvested pitaya fruit by enhancing the antioxidant potential of the AsA-GSH cycle and activating an antioxidant defense system to alleviate reactive oxygen species (ROS) accumulation.

Keywords: pitaya fruit; *p*-Anisaldehyde; quality; reactive oxygen species (ROS); AsA-GSH cycle; antioxidant activity

Citation: Xu, Y.; Cai, Z.; Ba, L.; Qin, Y.; Su, X.; Luo, D.; Shan, W.; Kuang, J.; Lu, W.; Li, L.; et al. Maintenance of Postharvest Quality and Reactive Oxygen Species Homeostasis of Pitaya Fruit by Essential Oil *p*-Anisaldehyde Treatment. *Foods* **2021**, *10*, 2434. https://doi.org/10.3390/foods10102434

Academic Editor: Victor Rodov

Received: 12 August 2021
Accepted: 10 October 2021
Published: 13 October 2021

1. Introduction

Pitaya fruit (*Hylocereus undatus*) is a tropical fruit originated from Latin America [1]. According to pulp and peel colour, pitaya fruit is classified into white flesh/yellow peel, white flesh/red peel, and red flesh/red peel fruits [2]. Owing to its desirable taste and texture, and abundant health-promoting compounds, the cultivation and consumption of pitaya have been growing substantially in the recent years [3]. Although pitaya is a non-climacteric fruit, it deteriorates and senesces rapidly after harvest due to the susceptibility to fungal diseases, and physiological disorders leading to shrinkage, thus limiting its storage and marketing potential [4,5]. Several reported treatments, such as cold storage [6], controlled atmosphere [7], plant hormone [8], X-ray irradiation [9] and synthetic chemicals [10], have been proved to control the postharvest diseases and fruit quality deterioration at varying degrees. Nevertheless, there is a continuing search for safer, low-cost, potent senescence inhibitors and antimicrobial technology to maintain quality of harvested pitaya fruit.

Essential oils are now increasingly used for the preservation of several fruits and vegetables due to its safety and antimicrobial properties. *p*-Anisaldehyde (PAA) (4-methoxybenzaldehyde) is a main component of the essential oil derived from seeds of *Pimpinella anisum* [11]. In laboratory media, fruit purees and fruit juices, PAA is confirmed to possess antimicrobial activities against a number of foodborne bacteria, such as *Bacillus subtilis*, *Pseudomonas aeruginosa*, *Listeria monocytogenes*, *Fusarium oxysporum*, and *Staphylococcus aureus*, yeasts (*Candida*) and mold strains (*Aspergillus niger*) [12]. Recently, *p*-Anisaldehyde/β-cyclodextrin combination as a fumigation agent effectively suppressed the growth of fungi in strawberry and preserved its storage quality [13]. It implies that PAA might be regulating postharvest physiological and biochemical behaviour of horticultural products. However, the potential of PAA on controlling postharvest deterioration of other postharvest fruits, and its underlying regulatory mechanisms, remains largely unknown.

Postharvest senescence and fruit quality deterioration involve metabolic disorder of reactive oxygen species (ROS) [14]. Overproduction of ROS, including superoxide anion radicals ($O_2^{\bullet -}$), hydrogen peroxide (H_2O_2), and hydroxyl radicals (OH^-) trigger oxidative damage to macromolecules, resulting in irreversible, deleterious changes in living cells [15]. ROS production is interlinked with ROS scavengers which encompass ROS enzymatic and non-enzymatic systems [16]. Enzymatic scavengers mainly include superoxide dismutase (SOD), peroxidase (POD), catalase (CAT), and the following enzymes involved in ascorbic acid-glutathione (AsA-GSH) cycle: ascorbate peroxidase (APX), glutathione reductase (GR), monodehydroascorbate reductase (MDHAR), and dehydroascorbate reductase (DHAR) [17]. The non-enzymatic scavenger system includes AsA, GSH, α-tocopherol, flavonoids, carotenoids and proline [18]. Mounting evidence from several decades indicates that excessive ROS generation caused by the disruption in the ROS production-scavenging balance can damage the cellular membrane structure, accelerate cell death, and reduce storability of harvested fruits, such as table grape [19], winter jujube [20], blueberries [21], and mango [22]. On the contrary, postharvest treatments, such as near-freezing temperature [23], acidic oxidizing water [24], and essential oils [25] for harvested fruits are proved to retain higher capacity of antioxidant and ROS scavenging ability, which help reduce pathogen infection and maintain fruit quality. In this sense, ROS homeostasis may serve as a common regulatory mechanism for fruits to control senescence process and maintain fruit quality.

Thus, in this work, the changes in physio-chemical properties related to fruit quality, total phenolics and flavones contents, 2,2-diphenyl-1-picrylhydrazyl (DPPH)-free radical scavenging rate, ROS generation, activities of ROS-scavenging enzymes, and components in ASA-GSH cycle in postharvest pitaya that received PAA treatment during storage were investigated. This research aimed to determine the role of ROS metabolism in PAA-mediated maintenance of fruit quality in postharvest pitaya fruit, as well as to validate the effectiveness of PAA treatment as an eco-friendly, safe and promising preservation method for extending the shelf life of harvested pitaya fruits.

2. Materials and Methods

2.1. Materials and Treatments

Red flesh/red peel pitaya (*Hylocereus polyrhizus* cv. 'Guanhuahong') fruits were harvested at the mature stage (~35 d after flower anthesis) from a commerical orchard in Guangzhou, China, and they were transferred to the laboratory immediately. Fruit with uniform shape, colour, size and no physical injuries and disease symptoms were selected and divided randomly into two groups (210 fruits in each group) for the following treatments.

The specific treatment procedures were conducted as follows: (1) PAA treatment- fruits were sprayed with 1 mM PAA solution until the PAA covered the fruit surface uniformly. PAA, at 1 mM, was chosen as the optimum concentration according to a preliminary experiment (data not shown). (2) Fruits evenly sprayed with distilled water served as control group. Thereafter, all treated fruits were air-dried, placed into a plastic box and stored at 20 °C for 15 d with 85–90% relative humidity.

Each treatment comprises three replicates, and samples of 10 pitaya fruits selected from each replicate were taken at Day 0 and at 3-day intervals for assessment of firmness and total soluble solids (TSS). Simultaneously, from the same samples flesh was collected and rapidly frozen at −80 °C for further analysis. For each parameter measurement, there were three replicates in each treatment at each time interval.

2.2. Determination of Fruit Physio-Chemical Quality

Every 10 fruits from each replicate were used for decay assessment. Decay incidence was measured based on the spoilage area with a scale composed of 0–5 degrees (0 = absence of decay; 1 =< 10% decay area; 2 = 10–25%; 3 = 25–50%; 4 = 50–75% and 5 => 75%), as described by Liu et al. [10]. The result of decay index was calculated by the equation:

$$\text{Decay incidence}\ (\%) = \frac{\sum(\text{decay scale} \times \text{number of fruit in each scale})}{(5 \times \text{total number of fruit})} \times 100 \quad (1)$$

Ten pitaya fruits per replication were weighed at Day 0 and at three-day interval during storage period. Weight loss was expressed as a percentage of weight lost compared to the initial weight.

Fruit firmness was measured at three equatorial points of the peeled fruit, using a GY-4 durometer equipment with a cylinder probe (12 mm diameter). The result was expressed as the N. TSS content was assessed by squeezing the fruit from the firmness test onto a digital refractometer (PAL-1, Atago, Japan) and was expressed as a percentage.

Betalain was extracted by homogenizing 0.5 g of sample with 5 mL of 80% methanol (*v/v*) solution by sonication for 10 min, and then centrifuged. Extraction was conducted twice. Betaxanthins and betacyanins were measured using a previously described method [26] through spectrophotometry at wavelengths of 478 nm and 538 nm, respectively. Content of both betalains was expressed as mg 100 g^{-1} of fresh weight (FW).

2.3. Measurement of Generation Rate of Superoxide Anion Radicals ($O_2^{\bullet-}$) and Hydrogen Peroxide (H_2O_2) Concentration

Production rate of $O_2^{\bullet-}$ and H_2O_2 content in pitaya pulp were determined using a kit (Comin, Suzhou, China), following the procedures of manufacturer's instructions. $NaNO_2$ was used as the standard for calculating the generation rate of $O_2^{\bullet-}$, which was expressed as nmol g^{-1} min^{-1} FW. H_2O_2 content was calculated with a standard curve constructed by H_2O_2, and expressed as μmol g^{-1} FW.

2.4. Assessment of Activity of Superoxide Dismutase (SOD), Peroxidase (POD), and Catalase (CAT)

Activity of SOD, POD, and CAT was determined using the biochemical kit (Comin, Suzhou, China) following the guidelines of manufacturer. The activity of these enzymes was expressed as unit (U) g^{-1} FW.

2.5. Determination of Components in Ascorbic Acid-Glutathione (ASA-GSH) Cycle

The metabolites in the ASA-GSH cycle mainly include AsA, dehydroascorbate (DHA), GSH and oxidized glutathione (GSSG), and their contents were determined according to the methods reported previously [27]. Content of ASA and DHA was calculated using ASA as a standard and were expressed as nmolg^{-1} FW. GSH and GSSG contents were calculated based on a standard curve of GSH and GSSG, respectively. The results of GSH and GSSG were expressed as μmol g^{-1} FW.

Activity of ascorbate peroxidase (APX), glutathione reductase (GR), monodehydroascorbate reductase (MDHAR), and dehydroascorbate reductase (DHAR) was measured using the reported methods [28]. The activity of all these enzymes was expressed as U g^{-1} FW.

2.6. Determination of Content of Total Phenolics, Flavonoids, and Scavenging Rate of DPPH Radical

Total phenolics and flavonoids contents were measured in accordance with the procedure as described by Han et al. [29]. The total phenolics content was calculated using the gallic acid as the standard, and result was expressed as mg of gallic acid equivalents (GAE) per gram of fresh weight (mg g^{-1} FW). The total flavonoids content was expressed as mg of rutin equivalent per gram of fresh weight (mg g^{-1} FW).

The scavenging rate of DPPH radical was determined by a biochemical kit (Comin, Suzhou, China). The absorbance of the reaction system at 515 nm was determined, and the result was finally expressed in percentage terms.

2.7. Gene Expression Analyses of Antioxidant Enzymes

Total RNA of pitaya fruit was extracted with EASYspin Plus Plant RNA kit (Aidlab Biotech, Beijing, China), following the manufacturer's instruction. Hifair™II 1st Strand cDNA Synthesis Super Mix for qPCR and Hieff® qPCR SYBR Green Master Mix (No Rox) (YEASEN Biotech, Shanghai, China) were employed to synthesize cDNA and to preform quantitative real-time PCR (qRT-PCR), respectively. *HpActin1* was selected as the internal control [30]. Gene expression was expressed relative to the expression level of *HpActin*1. The primers used in this study are listed in Supplementary Table S1.

2.8. Statistical Analysis

All data presented are means ± standard error of three biological replicates and were subjected to analysis of variance (ANOVA) using SPSS software. Mean values were compared using a Duncan's test to the significance level ($p < 0.05$ or $p < 0.01$).

3. Results

3.1. Effects of PAA on Visual Appearance and Physio-Chemical Quality Properties of Pitaya Fruits during Storage

Visual appearance of pitaya fruits in both groups almost remained unchanged in the initial six days of storage (Figure 1), but shrinkage of bracts and peel, and slight decay symptoms were observed in control fruit on Day 9. Further observations showed that decay, bract degreening, and water loss were more evident in control fruits compared with their counterparts in PAA treatment after 12 d of storage. Comparatively, PAA application maintained better freshness and appearance. On Day 15, the whole fruits decayed extensively in the control, while PAA treatment considerably delayed fruit decay.

Decay of pitaya fruit was significantly inhibited by PAA treatment (Figure 2A). The decay index was reduced from 70.66% in control to 44.67% in PAA-treated fruit (Figure 2A). Moreover, as shown in Figure 2B, fresh weight decreased throughout storage irrespective of treatment; however, the weight loss in the control group was more pronounced compared to that treated with PAA throughout the experiment. After 15 d of storage, weight loss of PAA-treated fruit was 26.85% lower ($p < 0.01$) than that of control fruit.

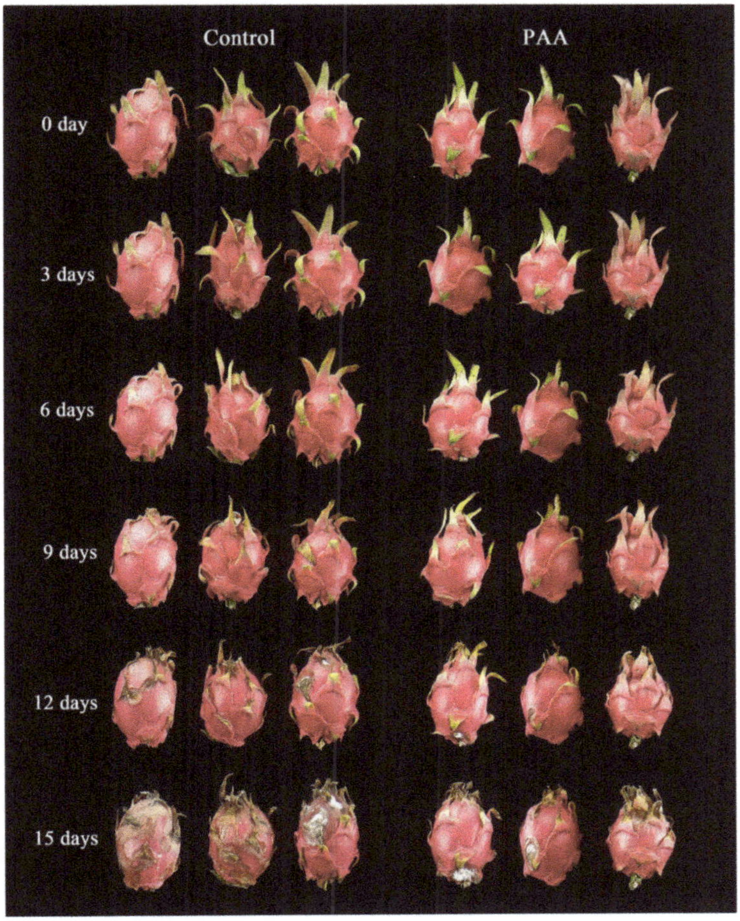

Figure 1. Changes in visual appearance during storage of pitaya fruits treated with PAA.

Figure 2C showed that fruit firmness of pitaya decreased continuously over the entire storage period. PAA treatment suppressed the loss of firmness during the entire storage. At the final storage time, fruits that were sprayed with PAA still retained firmer (8.24 N) than the control group (7.22 N). For TSS content, regardless of treatment, TSS content of pitaya fruit decreased linearly with storage time. Compared with the initial value (19.07%), TSS content in control pitaya fruits was decreased by 18.56% ($p < 0.05$) at the end of storage, while higher TSS content was observed in PAA-treated fruits throughout storage (Figure 2D).

The contents of betacyanins in postharvest pitaya fruits gradually increased until 9 days after treatment while it decreased over the rest of storage time (Figure 2E). Although no statistically significant differences between two groups were found during the first 9 d, PAA treatment maintained the betacyanins content. On Day 12, the content of betacyanins in pitaya fruits treated with PAA was significantly higher than that of control group, which was 1.14-fold that of control ($p < 0.05$). A similar variation was observed for the betaxanthins content of PAA-treated and control fruit (Figure 2F). Betaxanthin contents in PAA-treated fruit reached the maximum at 11.47 mg 100 g^{-1} FW on Day 9, which was 11.90% higher than that in control fruits.

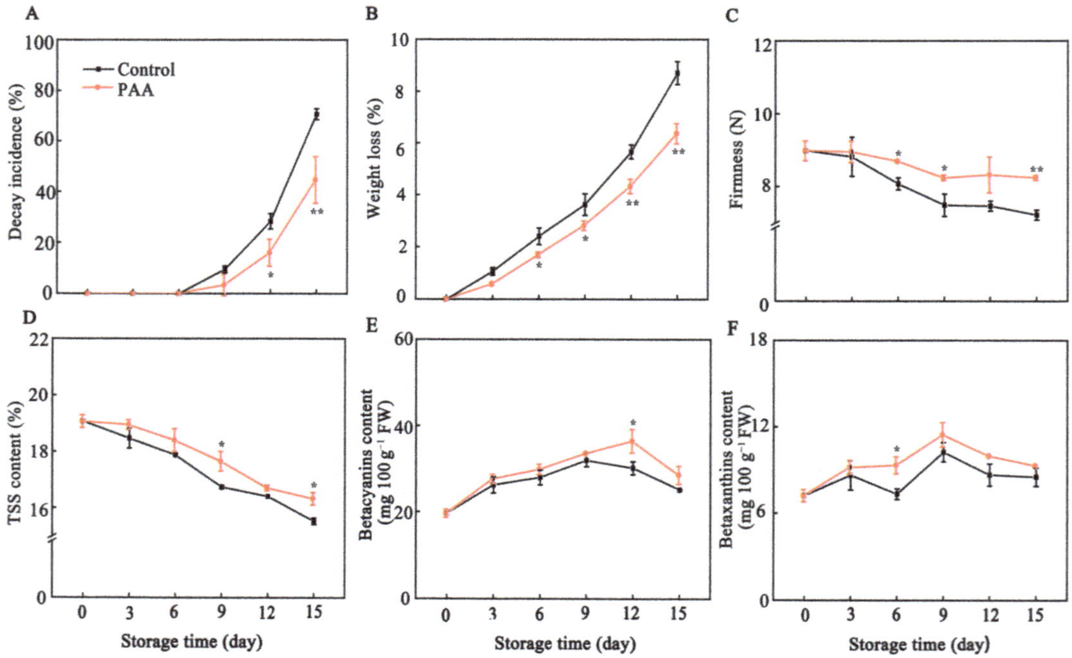

Figure 2. Changes in decay incidence (**A**), weight loss (**B**), firmness (**C**), total soluble solid (TSS) (**D**), betacyanins (**E**), and betaxanthins (**F**) during storage of pitaya fruits treated with PAA. Vertical bars represent the standard error of the mean. The asterisks indicated significant difference between two treatments during the same storage period (* $p < 0.05$, ** $p < 0.01$).

3.2. Effects of PAA on Generation Rate of $O_2^{\bullet -}$ and H_2O_2 Concentration of Pitaya Fruits during Storage

$O_2^{\bullet -}$ production rate in all both treatments significantly increased from storage Day 1 to Day 9, after which the levels were declined gradually until Day 15 (Figure 3A). However, the generation rate of $O_2^{\bullet -}$ in PAA-treated fruits was lower than that in control fruits throughout the storage.

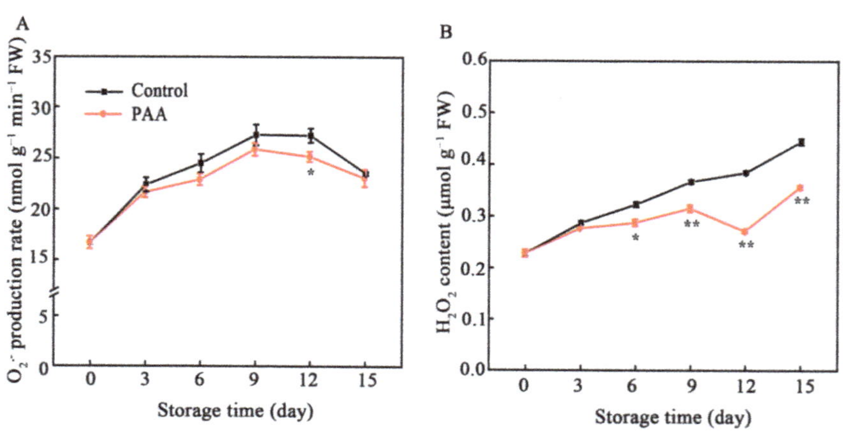

Figure 3. Changes in generation rate of $O_2^{\bullet -}$ (**A**) and H_2O_2 content (**B**) during storage of pitaya fruits treated with PAA. Vertical bars represent the standard error of the mean. The asterisks indicated significant difference between two treatments during the same storage period (* $p < 0.05$, ** $p < 0.01$).

H_2O_2 content increased in control fruits with storage time (Figure 3B). The accumulation of H_2O_2 in control fruits increased from an initial value of 0.23 μmol g^{-1} FW to a maximum of 0.44 μmol g^{-1} FW after 15 d of storage. PAA treatment significantly inhibited H_2O_2 production, in which the concentration of H_2O_2 on Day 12 was 19.74% lower than that in control fruits ($p < 0.01$).

3.3. Effects of PAA on POD, SOD and CAT Enzymatic Activity and Gene Expression in Pitaya Fruits during Storage

Activity of SOD and POD exhibited a similar trend, which rose considerably increase at early storage and dropped at the late storage period (Figure 4A,B). The maximum values of POD activity in PAA-treated fruits, and the SOD activity in both control and PAA treatment were all found on the third day. However, control group had the highest level on the twelfth day. Moreover, PAA treatment improved the activity of SOD and POD, with 28.00% and 28.53% higher ($p < 0.01$) than those of control pitaya fruits on Day 9, respectively. CAT activity in control fruits stayed at a stable low level during the whole storage. Until Day 15, CAT activity in PAA-treated fruits was 1.16 times than it was in control (Figure 4C).

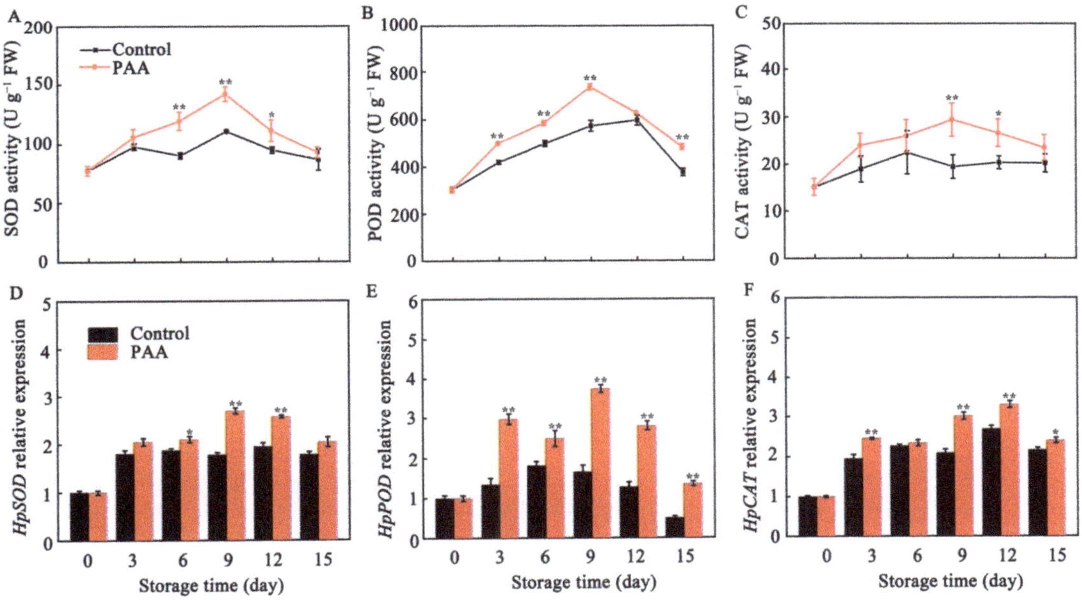

Figure 4. Changes in SOD activity (**A**), POD activity (**B**), CAT activity (**C**), *HpSOD* expression (**D**), *HpPOD* expression (**E**) and *HpCAT* expression (**F**) during storage of pitaya fruits treated with PAA. Vertical bars represent the standard error of the mean. The asterisks indicated significant difference between two treatments during the same storage period (* $p < 0.05$, ** $p < 0.01$).

As depicted in Figure 4D–F, there was a similar tendency between SOD, POD and CAT enzymatic activity and gene expression. The expression of *HpSOD*, *HpPOD* and *HpCAT* was obviously enhanced by PAA treatment, and a significant difference was found in the expression level of *HpPOD* throughout the storage.

3.4. Effects of PAA on Metabolite Content in ASA-GSH Cycle of Pitaya Fruits during Storage

As shown in Figure 5A,B, as storage time progressed, the contents of AsA and DHA in postharvest pitaya fruits peaked on Days 9 and 12, respectively, and then declined. AsA content in fruits treated with PAA was significantly higher than that of control fruits, however, DHA content in PAA-treated pitaya fruits was lower than that of control during the entire storage period. Application of PAA improved the ratio of AsA/DHA in pitaya fruits (Figure 5C). The ratio of AsA/DHA in PAA-treated pitaya fruits was 27.56% and 41.71% higher ($p < 0.01$) than that of control fruits on third and fifteenth day, respectively.

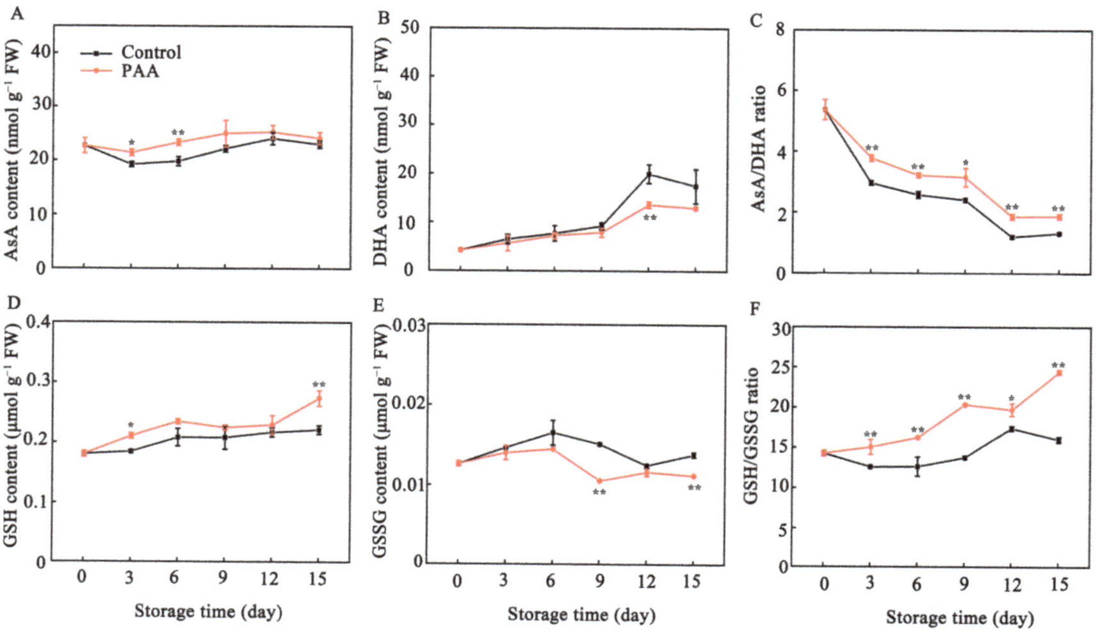

Figure 5. Changes in AsA content (**A**), DHA content (**B**), ratio of AsA/DHA (**C**), GSH content (**D**), GSSG content (**E**) and ratio of GSH/GSSG (**F**) during storage of pitaya fruits treated with PAA. Vertical bars represent the standard error of the mean. The asterisks indicated significant difference between two treatments during the same storage period (* $p < 0.05$, ** $p < 0.01$).

A gradual increase in GSH content was observed in both PAA-treated and control fruits (Figure 5D). Though values of GSH content in both groups showed no difference from storage Day 3 to Day 12, a higher level of GSH content was recorded in PAA-treated fruits during the whole storage. GSSG contents between PAA treatment and control group followed a similar trend, which increased slightly in the early storage period and declined afterwards (Figure 5E). The content of GSSG in PAA-treated pitaya was significantly lower than that in the control fruits on Days 9 and 15. Furthermore, the ratio of GSH/GSSG in pitaya was remarkably increased by PAA treatment compared with control (Figure 5F).

3.5. Effects of PAA on the Activity and Gene Expression of AsA-GSH Pathway Related Enzymes in Pitaya Fruits during Storage

As shown in Figure 6A, APX activity of pitaya fruits increased within the first 6 d, and fluctuated over the rest of storage, irrespective of treatment. PAA treatment resulted in significant increases in APX activity during most of storage. On Days 6 and 12, APX in PAA-treated fruits was 1.11 and 1.30 times higher ($p < 0.01$) than that of control, respectively. GR activity in both PAA-treated and control fruits increased steadily, and reached the maximum level on Day 12, and then declined for the remainder of storage (Figure 6B), but the rate of decline in PAA treatment during late storage was considerably less pronounced than that in the control. DHAR activity, which was higher at 3 d of storage, tended to decline during storage. However, significant differences ($p < 0.05$) of 13.32% and 17.86% over the controls were found after 6 d and 15 d of storage, respectively (Figure 6C). Furthermore, MDHAR activity fluctuated to a greater extent in pitaya fruits during storage (Figure 6D). In comparison to the control, MDHAR activity in the PAA-treated group was higher during the whole storage period, and the difference was extremely significant at 6 d, 9 d, and 12 d, which was 1.21-, 1.19- and 1.34- fold ($p < 0.01$) of control group, respectively.

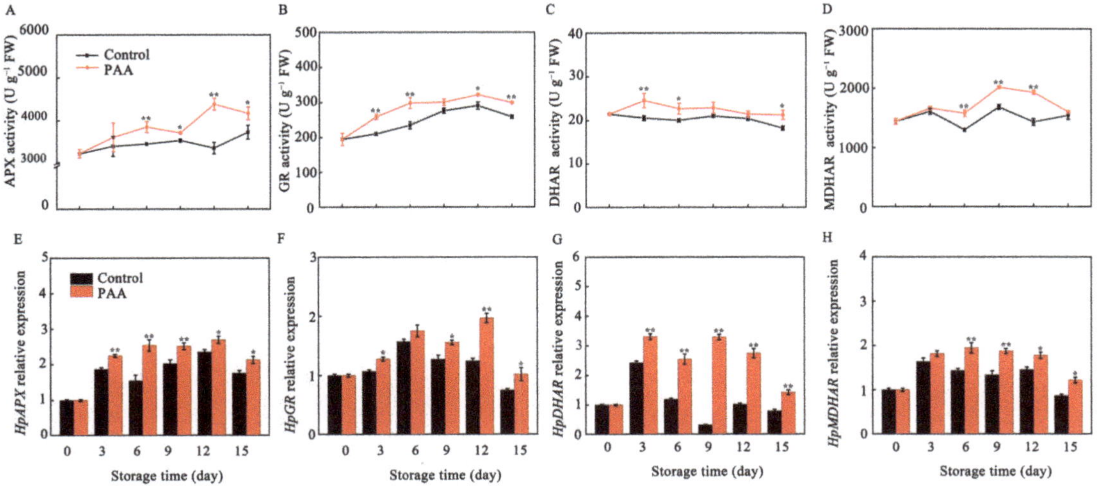

Figure 6. Changes in APX activity (**A**), GR activity (**B**), DHAR activity (**C**), MDHAR activity (**D**), *HpAPX* expression level (**E**), *HpGR* expression level (**F**), *HpDHAR* expression level (**G**) and *HpMDHAR* expression level (**H**) during storage of pitaya fruits treated with PAA. Vertical bars represent the standard error of the mean. The asterisks indicated significant difference between two treatments during the same storage period (* $p < 0.05$, ** $p < 0.01$).

The relative gene expression of *HpAPX*, *HpGR*, *HpDHAR*, *HpMDHAR* in postharvest pitaya fruits treated with PAA paralleled to those of corresponding enzymes activities. mRNA levels of *HpAPX* and *HpGR* tended to be up-regulated in the earlier storage period and down-regulated during the late storage period (Figure 6E,F). PAA treatment resulted in significantly higher expression of *HpAPX* and *HpGR* throughout the storage time. Similarly, expression of *HpDHAR* and *HpMDHAR* genes were up-regulated in PAA-treated pitaya fruits in comparison with control fruits (Figure 6G,H).

3.6. Effects of PAA on Content of Total Phenolics, Total Flavonoids and DPPH Radical-Scavenging Rate of Pitaya Fruits during Storage

The total phenolic and flavonoids content in control fruits was lower than that in the PAA-treated fruits over the storage period. Total phenolic in control samples declined from Day 9, whereas pitaya fruits in PAA treatment began to decrease from Day 12. Compared with untreated control, PAA-treated fruits showed 0.35- and 0.2-folds higher total phenolic and flavonoids after 15 d of cold storage, respectively (Figure 7A,B).

DPPH radical scavenging rate in both PAA-treated and control fruits during the experiment was shown in Figure 7C, with a persistent decline, except for values at 6 d. However, this decrease was suppressed by PAA treatment. On Day 15, the DPPH free radical scavenging rate of fruits under PAA treatment was 3.46% higher than that of the control.

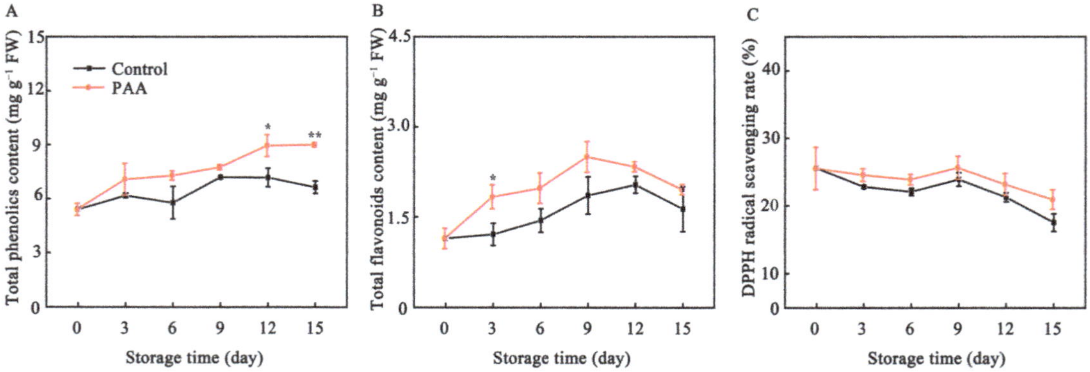

Figure 7. Changes in content of total phenolics (**A**), total flavonoids (**B**) and DPPH radical scavenging rate (**C**) during storage of pitaya fruits treated with PAA. Vertical bars represent the standard error of the mean. The asterisks indicated significant difference between two treatments during the same storage period (* $p < 0.05$, ** $p < 0.01$).

4. Discussion

Postharvest decay is a main limitation for the commercial value and storage life of pitaya fruit. With antimicrobial and insecticidal activity, essential oils are accepted as a prospective option for controlling postharvest fruit quality and safety [31]. The finding of the current study demonstrates that PAA treatment efficiently reduced the decay incidence of pitaya fruits (Figure 2A), which was consistent with the previous studies indicating that PAA could enhance resistance against disease development caused by green mold and blue mold of citrus fruits [32]. In addition, postharvest pitaya fruits undergo a loss of freshness which is characterized by a decline in bract greenness, increased weight loss, decreased fruit firmness and soluble solids [33]. In the present study, visual changes in skin and bract colour obviously varied between PAA treatment and control groups (Figure 1). Furthermore, the result here exhibited that PAA treatment efficiently reduced the weight loss, and delayed the decline in firmness, and TSS (Figure 2B–D) in pitaya fruits during storage. A little lower weight loss observed in PAA-treated fruit than that in the control fruit is possible due to the fact that the PAA functions as a coating agent on the surface of the pitaya fruit, impedes loss of moisture from the fruit. Virtually, weight loss is reportedly interlinked with respiration rate, thus, it is worth exploring the effect of PAA on fruit respiration in the future. As wilting incurred, the firmness of the fruit decreased during storage, while application of PAA maintained higher firmness, which both inhibited the rate of fruit softening and made the fruit less prone to mechanical and microbial damage. Conversely, Lin et al. reported that the fumigation using free PAA induced the loss of firmness, lightness of the surface color and cause a higher water loss [13]. A possible explanation for such opposite results is the differences in species and/or concentrations. Moreover, as a big reservoir of bioactive phytochemicals, pitaya contains betacyanins with remarkable pharmacological values [34]. In this study, a higher content of betacyanins and betaxanthins was retained in PAA-treated fruit as compared with control fruit (Figure 2E,F), which not only functioned as antioxidants but also contributed to maintenance of visual appearance, as the red colour of pitaya fruit is attributed to betacyanins. Therefore, these results suggest that PAA might suppress tissue decay and maintain the nutritional and flavour qualities of pitaya fruits.

According to the available reports, oxidative damage resulting from imbalance in both antioxidant response and ROS-generation affect fruit quality, fruit senescence and resistance to pathogens in most non-climacteric fruits [35]. SOD, POD, and CAT are most studied antioxidant enzymes. SOD dismutates $O_2^{\bullet -}$ to H_2O_2, representing the primary line of resistance against ROS. Then, POD and CAT act synergistically to disintegrate H_2O_2 into H_2O and O_2 [27]. Enhancing activity of antioxidant enzymes and their associated gene expression to modulate cellular redox homeostasis was previously shown to delay senescence and quality deterioration in various fruits. For example, Chen et al. indicated that enhanced activity of SOD, CAT and APX under 1-methylcyclopropene (1-MCP) treatment contributes to eliminating $O_2^{\bullet-}$ and maintaining the quality of pears [36]. Melatonin-induced fruit senescence inhibition has been shown to involve enhanced SOD, CAT, APX and POD activities [37]. Moreover, in pear up-regulation of *PcSOD* and *PcCAT* as well as enhanced activity of SOD and CAT reduced H_2O_2 production, leading to delayed senescence [38]. In the current study, ROS level in pitaya fruits increased as senescence progressed during storage (Figure 3). PAA markedly improved expression of *HpSOD*, *HpPOD* and *HpCAT* (Figure 4D–F) accompanied by increased activity of SOD, POD and CAT (Figure 4A–C) in harvested pitaya fruits during storage, which led to a lower level of $O_2^{\bullet-}$ and H_2O_2 content in PAA-treated pitaya fruits (Figure 3A,B). These findings indicate that the effect of PAA on reducing accumulation of ROS in pitaya fruits was correlated to the enhanced ROS-scavenging ability at both enzymatic and transcript levels, which, in turn, mitigated oxidative damage and the development of decay and senescence.

Out of the antioxidant enzymes, AsA and GSH have the direct capacity of quenching ROS. In addition, GSH participates in regeneration of AsA through AsA-GSH cycle to remove excess ROS [39]. In the AsA-GSH cycle, APX uses AsA as a substrate to catalyze the reduction of H_2O_2 to H_2O with concomitant production of MDHA, but owing to its unstable property, MDHA can dismutate into DHA or is regenerated into AsA through MDHAR, and DHA is further reduced to AsA by DHAR using reducing equivalents from GSH [40]. GR, a relevant component of the AsA-GSH cycle, catalyzes the conversion of GSSG to GSH form, allowing the maintenance of GSH/GSSG ratio [40]. The protective role of AsA and GSH as well as the ratio of AsA/DHA and GSH/GSSG in enhancing oxidant stress tolerance to delay senescence and maintain quality has been reported in several horticulture products [38,41,42]. Furthermore, given the importance of AsA-GSH cycle in antioxidant and stress resistance, key enzymes and genes involved in this cycle have also been extensively studied. Recently, Zhang et al. reported that 1-MCP induced expression of *AdAPX*, *AdDHAR* and *AdGR* but inhibited two isoforms of *AdMDHAR* expression, which was conducive to elevating the AsA content, scavenging H_2O_2, and postponing the senescence of kiwifruit [41]. A similar result was also found in bell pepper, where enhanced AsA-GSH cycle reduced H_2O_2, and $O_2^{\bullet-}$ content, overcoming the physiological disorders during cold storage [43]. Comparably, in our present study, the transcription of *HpAPX*, *HpGR*, *HpDHAR*, *HpMDHAR* was boosted by PAA treatment (Figure 6E–H), which was consistent with the PAA-enhanced activity of APX, GR, DHAR, and MDHAR (Figure 6A–D). Additionally, these enzymes together with higher ratio of AsA/DHA and GSH/GSSG, and lower DHA and GSSG contents (Figure 5A–F) explained the observed lower production of ROS under PAA treatment (Figure 3). These collectively indicate that the increased antioxidant capacity following PAA treatment, resulting in higher content of metabolites, enzyme activity and transcript abundance of genes involved in AsA-GSH cycle, play a vital role in ROS detoxification and redox state maintenance during postharvest storage of pitaya fruits.

In addition, total phenolics, and flavonoids as non-enzymatic antioxidant also fulfill a crucial role in protecting cells from oxidative damage. In plants, DPPH radical-scavenging capacity is generally used to evaluate the total non-enzymatic antioxidant capacity [44]. It has been reported that increasing DPPH radical-scavenging ability, total phenolics and flavonoids is positively correlated with the reduction of ROS and suppression of oxidative events enhanced in postharvest pitaya fruits treated with diphenyliodonium iodide [45], apple polyphenols [1], and methyl jasmonate [3]. In the present study, the decrease of DPPH radical scavenging rate was delayed by PAA treatment (Figure 7C), which was also accompanied by higher contents of total phenolic and flavonoids compared with control (Figure 7A,B), which partially help activate antioxidant responses and inhibit overproduction of ROS.

5. Conclusions

Based on above discussion, it is clear that postharvest application of PAA significantly dampened senescence and tissue decay, and effectively maintained the overall quality index of pitaya fruit. The enhanced postharvest disease resistance and quality preservation by PAA treatment might be associated with the reduction in ROS level and an increase in antioxidant capacity. The data suggest that this was attained through enhanced level of total phenolics, flavonoids, and DPPH radical scavenging, and increased gene expression and activity of SOD, POD and CAT as well as AsA-GSH cycle. The present research may help further elucidate the mechanism underpinning PAA-mediated preservation of postharvest pitaya fruit quality.

Supplementary Materials: The following are available online at https://www.mdpi.com/article/10.3390/foods10102434/s1, Table S1: primers sequence used for qRT-PCR in this study.

Author Contributions: Conceptualization, Y.Z., Z.C. and J.C.; methodology, Y.X. and Y.Z.; software, Y.X. and Y.Z.; validation, Y.X. and Y.Z.; investigation, Y.X. and Y.Z.; data curation, Y.X., Z.C., L.B., X.S., Y.Q., D.L., W.S., J.K., W.L., L.L., J.C. and Y.Z.; writing—original draft preparation, Y.X. and Y.Z.; writing—review and editing, Z.C. and J.C.; project administration, Y.Z., Z.C. and J.C.; funding acquisition, J.C., L.B. and Y.Q. All authors have read and agreed to the published version of the manuscript.

Funding: This research was funded by the Key Science and Technology Planning Project of Guangzhou (Grant No. 201904020015), National Natural Science Foundation of China (Grant No. 31801604), Science and Technology Program of Guangzhou (Grant No. 202002020060) and National Undergraduate Innovation and Entrepreneurship Training Program (Grant No. 202010564004).

Institutional Review Board Statement: Not applicable.

Informed Consent Statement: Not applicable.

Data Availability Statement: Not applicable.

Acknowledgments: The authors gratefully thank laboratory colleagues for their help. We are grateful to Prakash Lakshmanan (University of Queensland) for his critically reading and editing the manuscript.

Conflicts of Interest: The authors declare no conflict of interest.

References

1. Fan, P.H.; Huber, D.J.; Su, Z.H.; Hu, M.J.; Gao, Z.Y.; Li, M.; Shi, X.Q.; Zhang, Z.K. Effect ostharvest spray of apple polyphenols on the quality of fresh-cut red pitaya fruit during shelf life. *Food Chem.* **2018**, *243*, 19–25. [CrossRef]
2. Grimaldo-Juárez, O.; Terrazas, T.; Garcia-Velásquez, A.; Cruz-Villagas, M.; Ponce-Medina, J.F. Morphometric analysis of 21 pitahaya (*Hylocereus undatus*) genotypes. *J. Prof. Assoc. Cactus Dev.* **2007**, *9*, 99–117. [CrossRef]
3. Li, X.A.; Li, M.L.; Wang, L.; Wang, J.; Jin, P.; Zheng, Y.H. Methyl jasmonate primes defense responses against wounding stress and enhances phenolic accumulation in fresh-cut pitaya fruit. *Postharvest Biol. Technol.* **2018**, *145*, 101–107. [CrossRef]
4. Bellec, F.L.; Vaillant, F.; Imbert, E. Pitahaya (*Hylocereus* spp.): A new fruit crop, a market with a future. *Fruits* **2006**, *61*, 237–250. [CrossRef]
5. Zahid, N.; Ali, A.; Siddiqui, Y.; Maqbool, M. Efficacy of ethanolic extract of propolis in maintaining postharvest quality of dragon fruit during storage. *Postharvest Biol. Technol.* **2013**, *79*, 69–72. [CrossRef]
6. De Freitas, S.T.; Mitcham, E.J. Quality of pitaya fruit (*Hylocereus undatus*) as influenced by storage temperature and packaging. *Sci. Agric.* **2013**, *70*, 257–262. [CrossRef]
7. Ho, P.L.; Tran, D.T.; Hertog, M.; Nicola, B.M. Effect of controlled atmosphere storage on the quality attributes and volatile organic compounds profile of dragon fruit (*Hylocereus undatus*). *Postharvest Biol. Technol.* **2020**, *173*, 111406. [CrossRef]
8. Li, X.A.; Li, M.L.; Wang, J.; Wang, L.; Han, C.; Jin, P.; Zheng, Y.H. Methyl jasmonate enhances wound-induced phenolic accumulation in pitaya fruit by regulating sugar content and energy status. *Postharvest Biol. Technol.* **2018**, *137*, 106–112. [CrossRef]
9. Wall, M.M.; Khan, S.A. Postharvest quality of dragon fruit (*Hylocereus* spp.) after X-ray irradiation quarantine treatment. *HortScience* **2008**, *43*, 2115–2119. [CrossRef]
10. Liu, R.L.; Gao, H.Y.; Chen, H.J.; Fang, X.J.; Wu, W.J. Synergistic effect of 1-methylcyclopropene and carvacrol on preservation of red pitaya (*Hylocereus polyrhizus*). *Food Chem.* **2019**, *283*, 588–595. [CrossRef]
11. Shreaz, S.; Bhatia, R.; Khan, N.; Muralidhar, S.; Basir, S.F.; Manzoor, N.; Khan, L.A. Exposure of Candida to *p*-anisaldehyde inhibits its growth and ergosterol biosynthesis. *J. Gen. Appl. Microbiol.* **2011**, *57*, 129–136. [CrossRef]

12. Shi, C.; Zhao, X.C.; Meng, R.Z.; Liu, Z.J.; Zhang, G.N.; Guo, N. Synergistic antimicrobial effects of nisin and *p*-Anisaldehyde on *Staphylococcus aureus* in pasteurized milk. *LWT-Food Sci. Technol.* **2017**, *84*, 222–230. [CrossRef]
13. Lin, Y.; Huang, R.; Sun, X.X.; Yu, X.; Xiao, Y.; Wang, L.; Hu, W.Z.; Zhong, T. The *p*-Anisaldehyde/β-cyclodextrin inclusion complexes as fumigation agent for control of postharvest decay and quality of strawberry. *Food Control* **2021**, *130*, 108346. [CrossRef]
14. Lin, Y.F.; Chen, M.Y.; Lin, H.T.; Hung, Y.-C.; Lin, Y.X.; Chen, Y.H.; Wang, H.; Shi, J. DNP and ATP induced alteration in disease development of phomopsis longanae Chi-inoculated longan fruit by acting on energy status and reactive oxygen species production-scavenging system. *Food Chem.* **2017**, *228*, 497–505. [CrossRef] [PubMed]
15. Mittler, R. ROS are good. *Trends Plant Sci.* **2017**, *22*, 11–19. [CrossRef]
16. Choudhary, A.; Kumar, A.; Kaur, A. ROS and oxidative burst: Roots in plant development. *Plant Divers.* **2020**, *42*, 33–43. [CrossRef]
17. Ma, Y.Y.; Huang, D.D.; Chen, C.B.; Zhu, S.H.; Gao, J.G. Regulation of ascorbate-glutathione cycle in peaches via nitric oxide treatment during cold storage. *Sci. Hortic.* **2019**, *247*, 400–406. [CrossRef]
18. Kapoor, D.; Singh, S.; Kumar, V.; Romero, R.; Prasad, R.; Singh, J. Antioxidant enzymes regulation in plants in reference to reactive oxygen species (ROS) and reactive nitrogen species (RNS). *Plant Gene* **2019**, *19*, 100182. [CrossRef]
19. Zhang, Z.; Xu, J.; Chen, Y.; Wei, J.; Wu, B. Nitric oxide treatment maintains postharvest quality of table grapes by mitigation of oxidative damage. *Postharvest Biol. Technol.* **2019**, *152*, 9–18. [CrossRef]
20. Zhao, Y.T.; Zhu, X.; Hou, Y.Y.; Wang, X.Y.; Li, X.H. Postharvest nitric oxide treatment delays the senescence of winter jujube (*Zizyphus jujuba* Mill. cv. Dongzao) fruit during cold storage by regulating reactive oxygen species metabolism. *Sci. Hortic.* **2020**, *261*, 109009. [CrossRef]
21. Xu, F.X.; Wang, S.H.; Xu, J.; Liu, S.Y.; Li, G.D. Effects of combined aqueous chlorine dioxide and UV-C on shelf-life quality of blueberries. *Postharvest Biol. Technol.* **2016**, *117*, 125–131. [CrossRef]
22. Wannabussapawich, B.; Seraypheap, K. Effects of putrescine treatment on the quality attributes and antioxidant activities of 'Nam Dok Mai No.4' mango fruit during storage. *Sci. Hortic.* **2018**, *233*, 22–28. [CrossRef]
23. Zhao, H.D.; Liu, B.D.; Zhang, W.L.; Cao, J.K.; Jiang, W.B. Enhancement of quality and antioxidant metabolism of sweet cherry fruit by near-freezing temperature storage. *Postharvest Biol. Technol.* **2019**, *147*, 113–122. [CrossRef]
24. Chen, Y.H.; Hung, Y.-C.; Chen, M.Y.; Lin, M.S.; Lin, H.T. Enhanced storability of blueberries by acidic electrolyzed oxidizing water application may be mediated by regulating ROS metabolism. *Food Chem.* **2019**, *270*, 229–235. [CrossRef] [PubMed]
25. Taheri, A.; Behnamian, M.; Dezhsetan, S.; Karimirad, R. Shelf life extension of bell pepper by application of chitosan nanoparticles containing Heracleum persicum fruit essential oil. *Postharvest Biol. Technol.* **2020**, *170*, 111313. [CrossRef]
26. Xie, F.F.; Hua, Q.Z.; Chen, C.B.; Zhang, L.L.; Chen, J.Y.; Zhang, R.; Zhao, J.S.; Hu, G.B.; Zhao, J.T.; Qin, Y.H. Transcriptomics-based identification and characterization of glucosyltransferases involved in betalain biosynthesis in Hylocereus megalanthus. *Plant Physiol. Bioch.* **2020**, *152*, 112–124. [CrossRef] [PubMed]
27. Tan, X.L.; Zhao, Y.T.; Shan, W.; Kuang, J.F.; Lu, W.J.; Su, X.G.; Tao, N.G.; Lakshmanan, P.; Chen, J.Y. Melatonin delays leaf senescence of postharvest Chinese flowering cabbage through ROS homeostasis. *Food Res. Int.* **2020**, *138*, 109790. [CrossRef] [PubMed]
28. Yun, Z.; Gao, H.J.; Chen, X.; Chen, Z.S.Z.; Zhang, Z.Z.K.; Li, T.T.; Qu, H.X.; Jiang, Y.M. Effects of hydrogen water treatment on antioxidant system of litchi fruit during the pericarp browning. *Food Chem.* **2021**, *336*, 127618. [CrossRef]
29. Han, X.Y.; Mao, L.C.; Wei, X.P.; Lu, W.J. Stimulatory involvement of abscisic acid in wound suberization of postharvest kiwifruit. *Sci. Hortic.* **2017**, *224*, 244–250. [CrossRef]
30. Chen, C.B.; Wu, J.Y.; Hua, Q.Z.; Tel-Zur, N.; Xie, F.F.; Zhang, Z.K.; Chen, J.Y.; Zhang, R.; Hu, G.B.; Zhao, J.T.; et al. Identification of reliable reference genes for quantitative real-time PCR normalization in pitaya. *Plant Methods* **2019**, *15*, 70. [CrossRef]
31. Bakkali, F.; Averbeck, S.; Averbeck, D.; Idaomar, M. Biological effects of essential oils—A review. *Food Chem. Toxicol.* **2008**, *46*, 446–475. [CrossRef]
32. Che, J.X.; Chen, X.M.; Ouyang, Q.L.; Tao, N.G. *p*-Anisaldehyde exerts its antifungal activity against *Penicillium digitatum* and *Penicillium italicum* by disrupting the cell wall integrity and membrane permeability. *J. Microbiol. Biotechnol.* **2020**, *30*, 878–884. [CrossRef]
33. Chaemsanit, S.; Matan, N.; Matan, N. Effect of peppermint oil on the shelf-life of dragon fruit during storage. *Food Control* **2018**, *90*, 172–179. [CrossRef]
34. Jiang, H.T.; Zhang, W.; Li, X.X.; Shu, C.; Jiang, W.B.; Cao, J.K. Nutrition, phytochemical profile, bioactivities and applications in food industry of pitaya (*Hylocereus spp.*) peels: A comprehensive review. *Trends Food Sci. Technol.* **2021**, *116*, 199–217. [CrossRef]
35. Tian, S.P.; Qin, G.Z.; Li, B.Q. Reactive oxygen species involved in regulating fruit senescence and fungal pathogenicity. *Plant Mol. Biol.* **2013**, *82*, 593–602. [CrossRef]
36. Chen, Y.H.; Lin, H.T.; Shi, J.; Zhang, S.; Lin, Y.F.; Lin, T. Effects of a feasible 1-methylcyclopropene postharvest treatment on senescence and quality maintenance of harvested Huanghua pears during storage at ambient temperature. *LWT-Food Sci. Technol.* **2015**, *64*, 6–13. [CrossRef]
37. Wang, S.Y.; Shi, X.C.; Wang, R.; Wang, H.L.; Liu, F.Q.; Laborda, P. Melatonin in fruit production and postharvest preservation: A review. *Food Chem.* **2020**, *320*, 126642. [CrossRef] [PubMed]

38. Li, X.; Li, C.Y.; Cheng, Y.; Hou, J.B.; Zhu, J.H.; Ge, Y.H. Postharvest application of acibenzolar-S-methyl delays the senescence of pear fruit by regulating reactive oxygen species and fatty acid metabolism. *J. Agric. Food Chem.* **2020**, *68*, 4991–4999. [CrossRef] [PubMed]
39. Soares, C.; Carvalho, M.E.A.; Azevedo, R.A.; Fidalgo, F. Plants facing oxidative challenges-a little help from the antioxidant networks. *Environ. Exp. Bot.* **2019**, *161*, 4–25. [CrossRef]
40. Sharma, P.; Jha, A.B.; Dubey, R.S.; Pessarakli, M. Reactive oxygen species, oxidative damage, and antioxidative defense mechanism in plants under stressful conditions. *J. Bot.* **2012**, *161*, 217037. [CrossRef]
41. Zhang, Y.; Wang, K.; Xiao, X.; Cao, S.F.; Chen, W.; Yang, Z.F.; Shi, L.Y. Effect of 1-MCP on the regulation processes involved in ascorbate metabolism in kiwifruit. *Postharvest Biol. Technol.* **2021**, *179*, 111563. [CrossRef]
42. Li, C.Y.; Wei, M.L.; Ge, Y.H.; Zhao, J.R.; Chen, Y.R.; Hou, J.B.; Chen, Y.; Cheng, J.X.; Li, J.R. The role of glucose-6-phosphate dehydrogenase in reactive oxygen species metabolism in apple exocarp induced by acibenzolar-s-methyl. *Food Chem.* **2019**, *308*, 125663. [CrossRef] [PubMed]
43. Yao, M.Y.; Ge, W.Y.; Zhou, Q.; Zhou, X.; Luo, M.L.; Zhao, Y.B.; Wei, B.D.; Ji, S.J. Exogenous glutathione alleviates chilling injury in postharvest bell pepper by modulating the ascorbate-glutathione (AsA-GSH) cycle. *Food Chem.* **2021**, *352*, 129458. [CrossRef] [PubMed]
44. Zhang, D.D.; Xu, X.F.; Zhang, Z.K.; Jiang, G.X.; Feng, L.Y.; Duan, X.W.; Jiang, Y.M. 6-Benzylaminopurine improves the quality of harvested litchi fruit. *Postharvest Biol. Technol.* **2018**, *143*, 137–142. [CrossRef]
45. Li, X.A.; Li, M.L.; Han, C.; Jin, P.; Zheng, Y.H. Increased temperature elicits higher phenolic accumulation in fresh-cut pitaya fruit. *Postharvest Biol. Technol.* **2017**, *129*, 90–96. [CrossRef]

foods

MDPI

Article

Use of Acetic Acid to Partially Replace Lactic Acid for Decontamination against *Escherichia coli* O157:H7 in Fresh Produce and Mechanism of Action

Jiayi Wang [1,*], Yue Lei [2], Yougui Yu [1], Lebin Yin [1] and Yangyang Zhang [1]

[1] College of Food and Chemical Engineering, Shaoyang University, Shaoyang 422000, China; yufly225@163.com (Y.Y.); yinlebin0451@163.com (L.Y.); zy17680464475@163.com (Y.Z.)
[2] Institute of Rice Research, Guizhou Academy of Agricultural, Guiyang 550009, China; leiyue0917@163.com
* Correspondence: jiayiwangsyau@syau.edu.cn; Tel.: +86-138-8924-0025

Abstract: *Escherichia coli* O157:H7 is frequently detected in ready-to-eat produce and causes serious food-borne diseases. The decontamination efficacy of lactic acid (LA) is clearly established. In this study, LA was mixed with acetic acid (AA) to reduce costs while achieving consistent or better inhibitory effects. Time-kill curves and inoculation experiments using fresh-cut spinach and arugula indicated that 0.8%LA+0.2%AA shows similar antibacterial effects to those of 1%LA. To determine whether 1%LA and 0.8%LA+0.2%AA exert antibacterial effects by similar mechanisms, proteomics analysis was used. The proteins related to macromolecule localization, cellular localization, and protein unfolding were uniquely altered after the treatment with 1%LA, and the proteins related to taxis, response to stress, catabolic process, and the regulation of molecular function were uniquely altered after the treatment with 0.8%LA+0.2%AA. Based on these findings, combined with the results of a network clustering analysis, we speculate that cell membrane damage is greater in response to LA than to 0.8%LA+0.2%AA. This prediction was supported by cell membrane permeability experiments (analyses of protein, nucleotide, ATP, and alkaline phosphatase leakage), which showed that LA causes greater membrane damage than 0.8%LA+0.2%AA. These results provide a theoretical basis for the application of an acid mixture to replace LA for produce decontamination.

Keywords: acetic acid; decontamination; lactic acid

Citation: Wang, J.; Lei, Y.; Yu, Y.; Yin, L.; Zhang, Y. Use of Acetic Acid to Partially Replace Lactic Acid for Decontamination against *Escherichia coli* O157:H7 in Fresh Produce and Mechanism of Action. *Foods* **2021**, *10*, 2406. https://doi.org/10.3390/foods10102406

Academic Editor: Marios Mataragas

Received: 6 September 2021
Accepted: 7 October 2021
Published: 11 October 2021

1. Introduction

Fruits and vegetables are rich in essential vitamins, minerals, and fiber [1]. The FDA recommends the daily intake of three to five different vegetables and two to four different fruits [2]. With the acceleration of daily activities, demand for ready-to-eat fruits and vegetables has increased. However, because they are not cooked at high temperatures, there is a higher risk of diseases caused by foodborne pathogens [3]. *Escherichia coli* O157:H7 is often detected in ready-to-eat foods, particularly in fresh-cut vegetables [4]. According to a report from the United States Centers for Disease Control (CDC), in 2018–2020, there were 539 cases of infection by *Escherichia coli* O157:H7, of which 489 cases were related to leafy greens. The most serious events occurred in 2018, when romaine lettuce from the Yuma growing region caused 210 infections in 36 states, including 96 hospitalizations, 27 patients who developed a type of kidney failure called hemolytic uremic syndrome, and five deaths in Arkansas, California, Minnesota, and New York (CDC: Reports of Selected *E. coli* Outbreak Investigations) [5]. Therefore, decontamination is an effective method to ensure the safety of ready-to-eat vegetables.

For fresh fruits and vegetables, non-thermal technologies have shown antibacterial effects without affecting the quality. For example, bacteriophages, cold plasma, and pulsed light technologies have been effectively applied to fresh produce [6,7]. However, the technical equipment is relatively expensive, and the methods are not suitable for the large-scale decontamination of fresh produce. Chemical decontamination methods benefit from a low cost and moderate efficacy; among the established chemical sanitizers, chlorine-based agents have the lowest cost and moderate efficacy, and the concentration of free chlorine was recommended as 10–200 mg/L [6,8,9]. Recent research on chlorine sanitizers is focused on the prevention of the cross-contamination of the washing water [10–13]. However, chlorine sanitizers were criticized because they generate carcinogenic byproducts (e.g., trihalomethanes, haloacetic acids, haloketones, and chloropicrin) and form chlorate during fresh produce decontamination [6,14,15].

As another type of sanitizer, most organic acids are listed as generally recognized as safe (GRAS) by the FDA, and some studies have indicated that they have a higher efficacy than that of chlorine [16–18]. Lactic acid (LA) is the most widely used GRAS organic acid for fresh produce decontamination. However, lactic acid (LA) is relatively expensive. Among the GRAS organic acids (i.e., citric acid, acetic acid, lactic acid, malic acid, succinic acid, tartaric acid, and propionic acid), acetic acid (AA) has the lowest cost, which is about one-third that of LA [2].

In this study, the use of AA to partially replace LA for decontamination was evaluated. Fresh-cut baby spinach and arugula were selected as the models for the analyses of decontamination against *E. coli* O157:H7. Furthermore, the antibacterial activities of organic acids can be attributed to cellular anion accumulation, which is determined by the proportion of undissociated molecules. Compared with dissociated anions, undissociated acidic molecules have stronger lipophilicity, allowing them to penetrate the microbial cell membrane more easily. After penetration, the higher intracellular pH in the environment will promote acid molecule dissociation, and the dissociated anions will accumulate in the cell and exert toxic effects on DNA, RNA, and ATP synthesis [14,19] and promote acid-sensitive protein denaturation and changes in osmotic pressure [20]. We speculate that the effects of the acid mixture (AM) and LA are mediated by different underlying mechanisms, irrespective of the count reduction, due to the addition of dissociated acetate anions. Therefore, another objective of this work was to evaluate the difference in the mechanism underlying the effects of LA and AM using quantitative proteomic technology.

2. Materials and Methods

2.1. Time-Kill Curve Analysis

A single colony of *E. coli* O157:H7 (NCTC12900) was inoculated into nutrient broth (Hopebio, Qingdao, China) and cultured overnight at 37 °C. After adjusting the culture to 10^7–10^8 CFU/mL, 5 mL was centrifuged at 12,000× *g* for 10 min to obtain the cell pellet, followed by three washing steps using 0.85% NaCl solution. Then, the cells were resuspended in 1 mL of sterilized distilled water and supplemented with 4 mL of sanitizer to obtain the desired sanitizer concentration. The treatment groups were treated with 0.8%LA+0.2%AA, 0.6%LA+0.4%AA, 1%LA, and 1% AA, and the control group was treated with sterilized distilled water. After reaction for 0, 20, 40, 60, and 90 s, 1 mL of the above mixture was mixed with 5 mL of 0.04 M $K_2HPO_4 \cdot 3H_2O$ to neutralize the sanitizer [21].

After serial dilution, the suspension (0.1 mL) was surface-plated on modified sorbitol MacConkey agar (Hopebio, Qingdao, China) for the quantification of *E. coli* O157:H7.

2.2. Sample Preparation

Baby spinach and arugula were purchased from Microgreens (Guangzhou, China). After removing the stem and broken leaves, the remaining parts were rinsed for 30 s to remove dirt. The obtained samples were drained using a manual salad spinner sterilized with 75% ethanol.

2.3. Inoculation

The inoculation procedure followed our previously described methods [3,14]. A single colony of *E. coli* O157:H7 was inoculated into nutrient broth and cultured overnight at 37 °C. After adjusting the suspension to 10^9 CFU/mL, 0.5 or 5 mL was added into two stomacher bags containing 200 mL of sterilized 0.85% NaCl solution. Then, 10 g of sample was added to the bag and massaged for 20 min. The sample was then placed on a sterilized plastic tray in a biosafety cabinet for air drying and stored at 4 °C for 12 h to ensure sufficient bacterial attachment. The resulting samples showed a low inoculation level (10^3–10^4 CFU/g) and high inoculation level (10^6–10^7 CFU/g).

2.4. Decontamination and Microbiological Analysis

Minimizing water consumption and wastewater discharge rates remain challenging in the food industry [8]; thus, a ratio of 1:20 (*w*/*v*) that was sufficient to cover fresh-cut baby spinach and arugula was selected in this study. Inoculated samples were added to the sanitizers (as described in Section 2.1) and disinfected for 1.5 min under shaking at 120 rpm. The samples were then washed with tap water for 15 s to remove the residual sanitizer. The samples were transferred to a polyethylene terephthalate box (18 × 13 × 4 cm), packaged using a polyvinyl chloride cling film (Nan Ya, Taiwan, China), and stored at 4 °C [19]. The samples were analyzed on days 3, 5, and 7 [14]. Fifteen grams of each sample were homogenized with 85 mL of sterilized 0.85% NaCl solution for 1.5 min in a stomacher bag [14]. The microbiological counts were obtained as described in Section 2.1.

2.5. Sensory Analysis

Sensory analysis was performed at the end of storage (day 7). Nine trained panelists (ages 24–39 years) from Shijiashike Co. Ltd. (Liaoyang, Liaoning, China) were invited to evaluate sensory color, flavor, and crispness. A 3-point scale method, as described by Wang et al. [14], was used for evaluation: 0 'dislike extremely, no characteristic of the product', 5 'neither like nor dislike, acceptability threshold', and 10 'like extremely, very good product characteristics'. The plates containing samples were marked on the bottom and reordered before analysis. During the evaluation, only one person was allowed into the room (equipped with a 40-W white light without windows) and was not allowed to communicate with another person after evaluation. Between each time analysis of the flavor, drinking water was used to rinse the mouth three times, and the next evaluation was performed after 30 s.

2.6. Protein, Nucleotide, ATP, and AKP Leakage Analysis

After disinfection (treatment for 20 s) and neutralization, as described in Section 2.1, cells were centrifugated at 12,000× *g* for 10 min. The supernatants were filtered through 0.22 μm filters. The protein and nucleotide concentration in the supernatants was measured by a micro protein assay [22] and at a wavelength of 260 nm [23], respectively. ATP concentration and alkaline phosphatase (AKP) activity in the supernatants was measured using test kit (Jiancheng, Nanjing, China).

2.7. Proteomic Analysis

2.7.1. Protein Preparation

After disinfection (treatment for 20 s) and neutralization as described in Section 2.1, the cells were collected from the neutralization fluid using a membrane filter (0.22 μm; Millipore, Darmstadt, Germany). The cells on the membrane were washed off using SDT buffer (4% SDS, 100 mM Tris–HCl, 100 mM DTT, pH 8.0), followed by grinding under liquid nitrogen. The samples were then placed in a 100 °C boiling water bath for 10 min, subjected to ultrasonic treatment for 5 min in an ice bath (25 W for 3 s at intervals of 7 s), added to a 100 °C boiling water bath for 5 min, and centrifuged at 14,000× *g* for 30 min. The samples were filtered in a ultrafiltration tube (0.22 μm), and the protein concentration was quantified using BCA Test Kit (Beyotime, Shanghai, China).

2.7.2. Protein Digestion and Peptide Labeling

Total protein from each sample was digested using filter-aided proteome preparation (FASP) method, as described by Wisniewski et al. [24]. The peptide mixture was labeled using the 10-plex™ Isobaric Mass Tagging Kit (Thermo Scientific, Waltham, MA, USA) according to the manufacturer's instructions.

2.7.3. Peptide Fractionation

To improve the peptide identification quality, fractionation was performed using the Dionex UltiMate3000 HPLC System (Thermo Fisher, Waltham, MA, USA). The Gemini-NX 4.6 × 150 mm column (3 μm, 110 Å) (00F-4453-E0; Phenomenex, Torrance, CA, USA) was used as the chromatographic column. Elution was performed at a flow rate of 400 μL/min with a gradient of 100% buffer A (10 mM ammonium acetate, pH 10.0) for 5 min, 0–40% buffer B (10 mM ammonium acetate in 90% ACN, pH 10.0) for 20 min, 40–100% buffer B for 7 min, and 0–100% buffer A for 8 min. After fractionation, 40 fractions were subjected to vacuum centrifugation, reconstituted into 10 parts, and freeze-dried. The samples were stored at −80 °C until LC–MS/MS.

2.7.4. LC–MS/MS

After equilibrating the Thermo Scientific separation column (75 μm × 25 cm, 5 μm, 100 Å, C18) with 95% buffer A (0.1% formic acid), the sample was automatically loaded on the Thermo Scientific EASY Trap Column (100 μm × 2 cm, 5 μm, 100 Å, C18) and then separated with a linear gradient: 5–28% buffer B (0.1% formic acid in acetonitrile) for 40 min; 28% to 90% buffer B for 2 min; 90% buffer B for 18 min. Orbitrap-ELite (Thermo Finnigan, San Jose, CA, USA) mass spectrometer was used for the analysis. The detailed mass spectrometry parameters were as follows: detection mode: positive ion; scan range of precursor ions: 350–2000 m/z; resolution of MS: 60,000 at m/z 200; AGC target: 1e6; Maximum IT for MS: 10 ms; number of scan ranges: 1; dynamic exclusion: 30 s; most intense signals for MS/MS: top 15; MS2 activation type: HCD; isolation window: 2 m/z; resolution of MS/MS: 15,000 at m/z 100; microscans: 1; maximum IT for MS/MS: 100 ms; AGC target: 5e4; normalized collision energy: 35 eV; underfill ratio: 0.1%.

2.7.5. Database Searching and Data Analysis

The raw data obtained as described in Section 2.7.4 were processed using Proteome Discover 2.3. A search for fragmentation spectra was performed using the Mascot search engine embedded in Proteome Discoverer against the uniprot_*Escherichia_coli*_O157:H7. fasta database. The search parameters were as follows: Type of search: MS/MS Ion search; Enzyme: Trypsin; Mass Values: Monoisotopic; Max Missed Cleavages: 2; Fixed modifications: Carbamidomethyl (C), TMT 10plex (N-term), TMT 10plex (K); Variable modifications: Oxidation (M); Peptide Mass Tolerance: ±20 ppm; Instrument type: ESI-TRAP; Fragment Mass Tolerance: 0.1 Da; Protein Mass; Unrestricted; Decoy database pattern: Ture; Database: uniprot-*Escherichia_coli*_O157:H7.fasta. Peptide identification results were filtered against the standard for a false discovery rate of <1%. Peptide ion peak intensities were collected, and the median peptide ratio was calculated. Then, protein quantification data from each channel were processed by the median normalization method to obtain the final protein quantification results.

2.8. Bioinformatics and Statistical Analysis

A fold change of >1.2 and $p < 0.05$ were thresholds for the identification of differentially expressed proteins (DEPs) [25]. A Gene Ontology (GO) enrichment analysis was performed using Blast2GO (https://www.blast2go.com/ (accessed on 1 May 2021)) [26]. Protein–protein interaction networks were analyzed using STRING (http://string-db.org/ (accessed on 10 July 2021)) [27], and the interaction score was set to 0.7.

Data for sensory characteristics, microbial counts, protein, and nucleotide leakage were analyzed using SPSS 22.0. Differences in mean values were analyzed using Duncan's multiple range tests, and p-values of <0.05 were considered significant. All data were expressed as means ± standard deviations. All experiments were independently replicated three times. Three samples were taken on each sampling day and were analyzed in duplicate for a total of six analyses per replicate.

3. Results and Discussion

3.1. Time-Kill Curves

Before the fresh produce decontamination and proteomic analyses, it is necessary to determine the time-kill curves under pure culture conditions. We found that 1%AA had the weakest antibacterial effect. After disinfection for 20 s, 1%AA only resulted in a 0.35 log reduction of the *E. coli* O157:H7, and, at the end of the disinfection period (90 s), only a 0.73 log reduction was observed (Figure 1). When the LA was included, the antibacterial effect was improved. In particular, 0.6%LA+0.4%AA reduced the *E. coli* O157:H7 by 0.73 log after 20 s and by 2.81 log at 90 s. When the concentration of LA was increased, 0.8%LA+0.2%AA resulted in a 4.29 log reduction in the *E. coli* O157:H7 within 90 s and showed a similar effect to that of 1% LA. Since long-term disinfection may alter the expression of most proteins, short disinfection periods should be selected for further proteomics analysis [25,28]. Therefore, 0.6%LA+0.4%AA, 0.8%LA+0.2%AA, and 1%LA were selected as the treatment groups for subsequent analyses of the decontamination of fresh produce, and 0.8%LA+0.2%AA and 1%LA for 20 s were selected for a proteomic analysis.

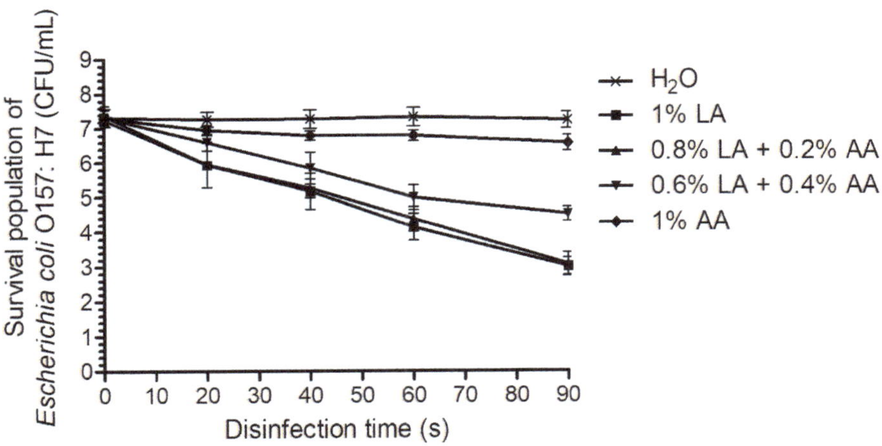

Figure 1. Time-kill curves for different combinations of sanitizers against *Escherichia coli* O157:H7. AA, acetic acid; LA, lactic acid.

3.2. Decontamination Efficacies of Various Combinations against Escherichia coli O157:H7 on Fresh-Cut Baby Spinach and Arugula

Owing to the complex and diverse contamination conditions for fresh-cut vegetables [29], two levels of contamination were evaluated in this study, low (10^3–10^4 CFU/g) and high (10^6–10^7 CFU/g). Under low contamination conditions, the *E. coli* O157:H7 on fresh-cut spinach and arugula increased significantly as the storage time increased (Figure 2a1,a2). The treatment with 1%LA and 0.8%LA+0.2%AA resulted in the lowest counts, with no significant difference between these two groups. From days 3 to 7, the effect of 0.6%LA+0.4%AA was significantly greater than those of the other two groups on fresh-cut arugula (Figure 2a2). For high contamination, the *E. coli* in the control group did not increase significantly from days 0 to 7, consistent with the results of previous studies [14,30], and the effects of 1%LA were similar to those of 0.8%LA+0.2%AA from days 0 to 7, and these two groups showed significantly lower *E. coli* O157:H7 counts than those of the 0.6% LA+0.4%AA group (Figure 2b1,b2). Counts on day 0 and day 7 differed significantly in the 0.6% LA+0.4%AA group but not in the other two groups (Figure 2b1,b2), indicating that 1%LA and 0.8%LA+0.2%AA are more suitable than 0.6%LA+0.4%AA for the decontamination of fresh-cut spinach and arugula with high *E. coli* O157:H7 contamination. In summary, irrespective of the contamination level, 1% LA and 0.8%LA+0.2%AA are recommended, consistent with the results of the time-kill curve analyses.

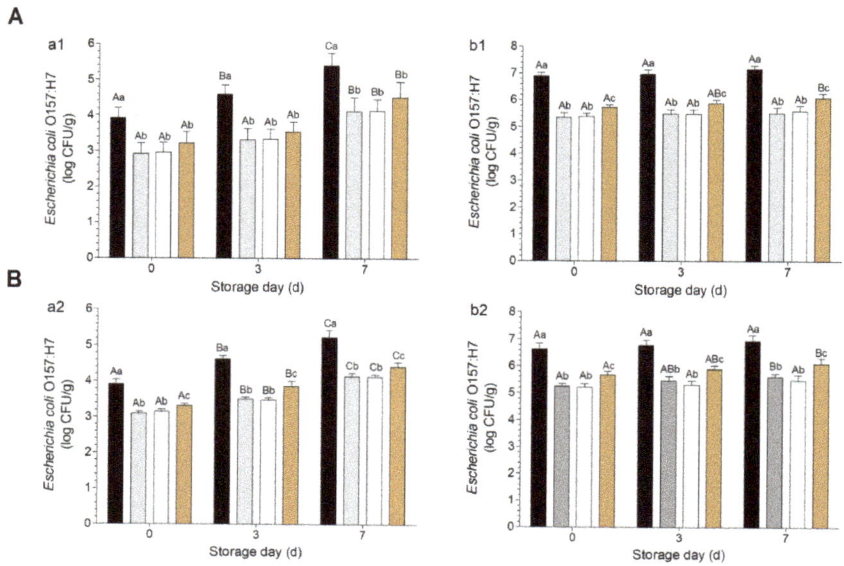

Figure 2. Decontamination effects of different combinations against *Escherichia coli* O157:H7 on fresh-cut spinach (**A**) and arugula (**B**). Panels a and b show results for low and high inoculation levels, respectively. Black, gray, white, and brown bars indicate the results for the control, 1% LA, 0.8% LA plus 0.2% AA, and 0.6% LA plus 0.4% AA, respectively. Note: within the same day, mean values with different lowercase letters are significantly different from each other ($p < 0.05$); within the same treatment, mean values with different capital letters are significantly different from each other ($p < 0.05$).

3.3. Effects of Sanitizers on Sensory Characteristics of Fresh-Cut Baby Spinach and Arugula

Decontamination will cause damage to the surface of fresh-cut vegetables [14,31]. If the damage is minor, fresh produce will repair itself [32]. If the damage is serious, it will cause a loss of flavor, browning, and water loss [33]. Damage to fresh-cut vegetables is not observed immediately; instead, it appears gradually during storage. In this study, the color, flavor, and crispness at the end of storage (7 d) were evaluated. The three treatments did not negatively affect the sensory qualities compared with those in the control group (Figure 3A,B). However, according to previous studies, AA concentrations exceeding 1% had negative effects on the quality of fresh-cut vegetables. For example, Wang et al. [34] observed browning blots on 1%AA-treated fresh-cut lettuce after storage for 5 d. Vijayakumar and Wolfhall [35] used 6%AA to disinfect fresh-cut lettuce and observed significant reductions in parameters related to appearance, taste, texture, and overall acceptance compared with lettuce treated with lemon juice, apple vinegar, and bleaching powder. When fresh-cut spinach and lettuce were stored for 7 d, the a * value in the AA treatment group was significantly higher than that of the control group [36].

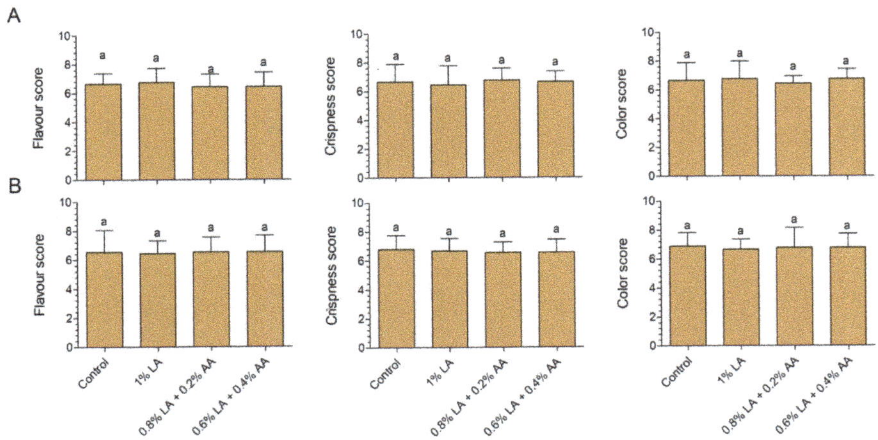

Figure 3. Effects of different combinations of sanitizers on the flavor, crispness, and color of fresh-cut baby spinach (**A**) and arugula (**B**). Different letters above the columns indicate significant differences ($p < 0.05$). AA, acetic acid; LA, lactic acid.

3.4. *Effects of Sanitizers on the Proteome*

Compared with gel-based proteomics, mass spectrometry-based proteomic analyses are now widely used owing to their high-throughput capacity, repeatability, and high success rate for protein identification [37]. In the present study, 15,376 peptides (Table S1) corresponding to 2430 proteins (Table S2) were successfully identified. There were 1755 DEPs (794 upregulated and 961 downregulated) in the LA–CK comparison and 1835 DEPs (761 upregulated and 1074 downregulated) in the AM–CK comparison, indicating that over 50% of the identified proteins were differentially expressed after 20 s of treatment. In AM–LA, 155 DEPs (65 upregulated and 90 downregulated) were found, with substantial overlap between the DEPs in the LA–CK and AM–CK comparisons, which may reflect the slight difference in the LA concentrations. Although the LA concentration differs by 0.2% between the groups, it is possible to achieve similar antibacterial effects via different biological processes.

3.5. *Venn Diagram Analysis*

We evaluated whether the mechanisms underlying the antibacterial effects differ between LA and AM. Therefore, DEPs enrichment for biological processes (BPs), which may be related to the observed antibacterial effects, were evaluated by a GO analysis and a Venn diagram. In total, 15 identical BPs were identified in the two comparisons (Figure 4). This result indicates that, although the difference in the LA concentration between the AM and LA groups is only 0.2%, the BPs affected by the treatment are similar. However, three and four unique BPs, as shown in Figure 4, were identified, indicating that the 0.2% difference in the LA results in changes in the BPs.

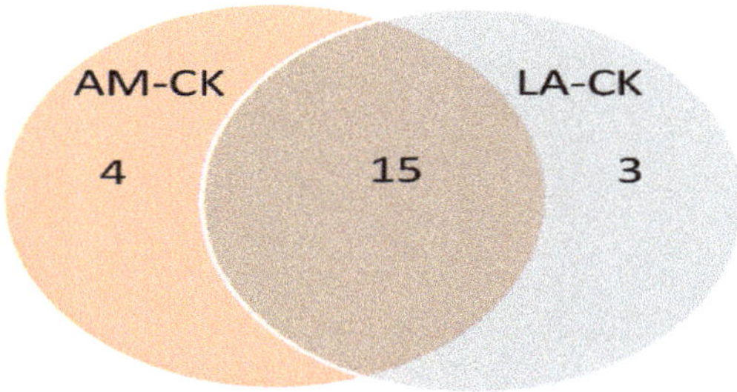

Figure 4. Distribution of biological processes (GO level 3) associated with differentially expressed proteins after treatment with AM and LA. LA: lactic acid; AM: acid mixture, 0.8% LA + 0.2% AA; CK: control.

3.6. Unique Enriched Biological Processes and Network Clustering of LA–CK

The three unique BPs in LA–CK were macromolecule localization, cellular localization, and protein unfolding, associated with 58, 32, and 5 DEPs, respectively. A protein–protein interaction analysis can be used to identify key functional clusters predicted to induce changes in the whole network [38–40], and the results indicate protein export (Figure S1A,B), outer membrane, and gram-negative porins (Figure S1A) were involved in the BPs macromolecule and cellular localization.

Protein export is the active transport of proteins from the cytoplasm to the exterior of the cell or to the periplasmic compartment in gram-negative bacteria. In this process, the Sec-dependent pathway, FtsY, and YidC (Table 1) are responsible for transporting newly synthesized proteins into or across the cell membrane [41,42].

Table 1. Related differentially expressed proteins in network clusters of LA–CK.

UniProt Accession Number	Gene ID	Protein Name	Fold Change
		Protein export	
P0AG88	secB	Protein-export protein SecB	2.35
P0AGA1	secG	Protein-export membrane protein SecG	2.06
P0AGA4	secY	Protein translocase subunit SecY	0.53
P0AG91	secD	Protein translocase subunit SecD	0.65
P0AG95	secF	Protein translocase subunit SecF	1.24
P65625	yidC	Membrane protein insertase YidC	0.43
P0AGD9	ffh	Signal recognition particle protein	0.38
		Outer membrane and gram-negative porins	
P61318	lolA	Outer-membrane lipoprotein carrier protein	0.82
P0ADC5	lolC	Lipoprotein-releasing system transmembrane protein LolC	0.37
P0ABV8	tolR	Tol-Pal system protein TolR	2.07
P0ABV0	tolQ	Tol-Pal system protein TolQ	0.46
Q8X8E2	lolE	Lipoprotein-releasing system transmembrane protein	0.45
P0AC04	bamD	Outer membrane protein assembly factor BamD	2.45
P0AEU9	skp	Chaperone protein Skp	2.25
P0A9V3	lptB	Lipopolysaccharide export system ATP-binding protein LptB	1.42

LA: lactic acid; CK: control.

Interestingly, another network cluster (i.e., outer membrane and gram-negative porin proteins; Table 1) shows a different location compared to the protein export proteins (typically located on the inner membrane). We hypothesize that 1%LA exerted greater cell membrane damage as compared with 1%AM. Information on the release of the cell constituents reveals the integrity of the cell membrane [43]. Protein, ATP, and nucleotide leakage from intracellular to extracellular spaces can reflect changes in the membrane integrity [31,43,44]. AKP activity in the extracellular environment can reflect the damage extent of the cell membrane. In this work, the leakage of protein, nucleotide, ATP, and AKP was significantly greater for 1% LA than 1% AM (Figure 5), consistent with our hypothesis.

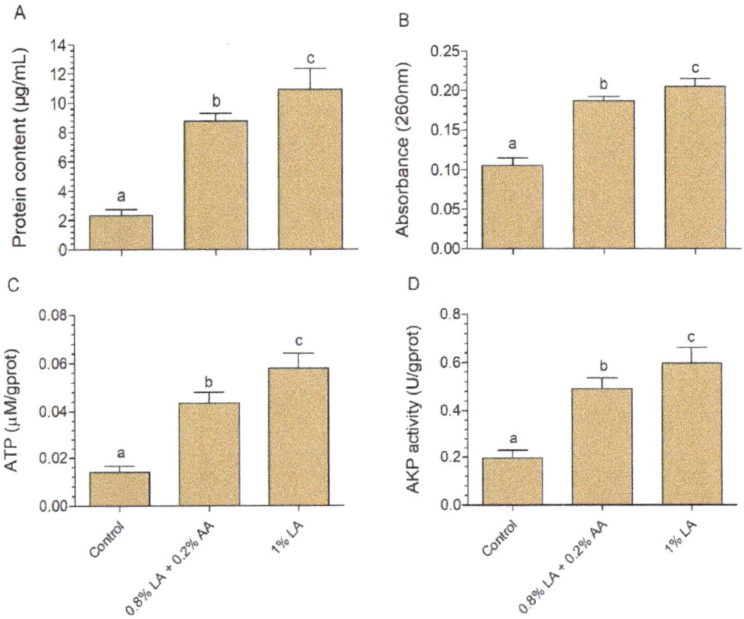

Figure 5. Effects of an acid mixture and lactic acid on protein content (**A**), nucleotide content (**B**), ATP content (**C**), and AKP activity (**D**) in *Escherichia coli* O157:H7. Different letters above the columns indicate significant differences ($p < 0.05$). AA, acetic acid; LA, lactic acid.

3.7. Unique Enriched Biological Processes and Network Clustering of AM–CK

The four unique BPs in AM–CK were taxis, response to stress, catabolic process, and the regulation of molecular function, associated with 18, 233, 203, and 35 DEPs, respectively. According to a protein interaction analysis, the phosphotransferase (PTS) system (Figure S2A) and glycerophospholipid metabolism (Figure S2A), flagellar assembly and bacterial chemotaxis (Figure S2B), and homologous recombination and DNA repair (Figure S3) were involved in the BPs catabolic process, taxis, and response to stress, respectively.

PTS (Table 2) is a distinct mechanism used by bacteria for sugar uptake, where the energy source is phosphoenolpyruvate, which are responsible for the *E. coli* O157:H7 sugar metabolism [45]. Similarly, another study [46] has shown that disodium succinoyl glycyrrhetinate, a derivative of glycyrrhetinic acid, suppresses the sugar metabolism in the cytoplasm at the protein level. The PTS system is multicomponent system that involves enzymes in the plasma membrane and the cytoplasm [47]. Moreover, the PTS system can be activated by acids, antibacterial agents, and salt [48].

Glycerophospholipid metabolism synthesizes the membrane components, and glycerophospholipid drives the formation of the lipid bilayer [49]. In this work, the proteins associated with glycerophospholipid metabolism were located in the cytoplasm and belonged to the major intrinsic protein family (Table 2), which includes transmembrane protein channels, e.g., aquaporins, aquaglyceroporins, and S-aquaporins [50]. According to a previous study, after microorganisms sense a drug, phospholipid synthesis is accelerated for drug transport [51,52], suggesting that the AM applied to the *E. coli* O157:H7 may induce glycerophospholipid metabolism in the cytoplasm.

Table 2. Related differentially expressed proteins in network clusters of AM–CK.

UniProt Accession Number	Gene ID	Protein Name	Fold Change
	Starch and sucrose metabolism, and phosphotransferase system		
Q8XDG9	mtlD	Mannitol-1-phosphate 5-dehydrogenase	1.27
Q8X677	manA	Mannose-6-phosphate isomerase	0.27
Q8XE22	pfkB	Phosphofructokinase	0.59
A0A0H3JCR2	pgm	Phosphoglucomutase	0.35
P0AB72	fbaA	Fructose-bisphosphate aldolase class 2	0.57
P62709	gpmA	2,3-bisphosphoglycerate-dependent phosphoglycerate mutase	2.34
P0A6V9	glk	Glucokinase	0.60
Q8XCE1	treC	Trehalose-6-P hydrolase	0.38
Q8X710	malQ	4-alpha-glucanotransferase	0.37
Q8X6Y1	glgP	Alpha-1,4 glucan phosphorylase	0.41
Q8X6X8	glgX	Glycogen debranching enzyme	0.34
	Glycerophospholipid metabolism and major intrinsic proteins		
P0A997	glpC	Anaerobic glycerol-3-phosphate dehydrogenase subunit C	0.59
P0A6F4	glpK	Glycerol kinase	0.55
P0A9C1	glpA	Anaerobic glycerol-3-phosphate dehydrogenase subunit A	0.69
A0A0H3JI74	glpD	Glycerol-3-phosphate dehydrogenase	0.60
P0A6S9	gpsA	Glycerol-3-phosphate dehydrogenase [NAD(P)+]	0.38
	Flagellar assembly and bacterial chemotaxis		
P0A966	cheW	Chemotaxis protein CheW	2.83
P0AE68	cheY	Chemotaxis protein CheY	2.80
Q8XCF9	cheB	Protein-glutamate methylesterase/protein-glutamine glutaminase	0.60
P0ABZ3	fliG	Flagellar motor switch protein FliG	1.67
P0ABY0	fliL	Flagellar protein FliL	2.00
	Homologous recombination and DNA repair		
Q8X8H1	polA	DNA polymerase I	0.78
P0A7G8	recA	Protein RecA	1.41
P0A7H2	recF	DNA replication and repair protein RecF	0.67
Q8XDN4	mutL	DNA mismatch repair protein MutL	0.40
Q8X8P5	uvrD	DNA helicase	0.58
P0A811	ruvA	Holliday junction ATP-dependent DNA helicase RuvA	0.69
Q8X5H9	ftsK	DNA translocase FtsK	0.21
P0A7C4	lexA	LexA repressor	2.07

AM: acid mixture; CK: control.

Homologous recombination and DNA repair (Table 2) are involved in the response to stress and particularly the SOS response. The LexA repressor negatively regulates the SOS genes. Once the pool of LexA decreases, the repression of the SOS genes decreases [53]. In this work, the LexA repressor was upregulated 2.02-fold, indicating that DNA repair was stimulated and that AM may exert more damage against *E. coli* O157:H7 DNA as compared with LA.

Regarding the proteins involved in flagellar assembly and bacterial chemotaxis (Table 2), FliG and FliL are flagellar motor switch proteins and interact with FliM to control the direction of the *E. coli* movement [54]. CheY and CheW are responsible for transmitting the signal obtained from the chemoreceptor to the flagellar rotator in response to an external stimulus [55]. The upregulation of these proteins (Table 2) indicated that AM may exert greater chemical stress than LA. In a previous study [56], oil isolated from fingered citron promoted chemotaxis and flagella assembly in *Listeria monocytogenes* at the transcriptomic level.

4. Conclusions

The effects of LA and AM on *E. coli* O157:H7 on fresh-cut produce as well as the mode of action at a proteome level were evaluated. A produce decontamination experiment showed that LA and AM have similar decontamination efficacies against the *E. coli*

O157:H7 on fresh-cut spinach and arugula, consistent with the results of the time-kill curves. In addition, the AM and LA did not negatively affect the sensory characteristics of the fresh-cut produce after storage. Accordingly, AM is a cheaper alternative to LA for the decontamination of fresh-cut produce. At the proteome level, LA and AM affected distinct biological processes. In particular, macromolecule localization, cellular localization, and protein unfolding were uniquely affected by LA, and taxis, response to stress, catabolic process, and the regulation of molecular function were uniquely changed after the treatment with AM. Further network clustering and cellular component analyses showed that membrane damage that may be induced by LA is greater than that induced by AM. Additionally, AM may have a stronger inhibitory effect on the biological processes in the cytoplasm, including DNA repair, bacterial chemotaxis, sucrose metabolism, and glycerophospholipid metabolism. Cell membrane permeability experiments confirmed that LA has a stronger damaging effect on cell membranes than AM. In addition, *E. coli* O157:H7 is a hazard for fresh meat, and lactic acid is used in various countries to decontaminate fresh meat. Whether the decontamination effect of AM is consistent with that of LA should also be evaluated in future studies. In addition, the relationship between decontamination and ecological changes should be further evaluated, including in-depth 16S rRNA and metatranscriptomic studies.

Supplementary Materials: The following are available online at https://www.mdpi.com/article/10.3390/foods10102406/s1, Figure S1: Protein–protein interaction network for differentially expressed proteins involved in macromolecule localization (A) and cellular localization (B). Yellow points in A indicate the network cluster related to cell outer membrane and gram-negative porin; red points in A and B indicate the network cluster related to protein export, Figure S2: Protein–protein interaction network for differentially expressed proteins involved in catabolic process (A) and taxis (B). Blue points in A indicate the network cluster related to starch and sucrose metabolism, and phosphotransferase system; red points in A indicate the network cluster related to glycerophospholipid metabolism and major intrinsic protein; red points in B indicate the network cluster related to flagellar assembly and bacterial chemotaxis, Figure S3: Protein–protein interaction network for differentially expressed proteins involved in the response to stress. Red points indicate the network cluster related to homologous recombination and DNA repair, Table S1: Peptide information for identified proteins, Table S2: Information on the identified proteins.

Author Contributions: Conceptualization, J.W., Y.L. and Y.Z.; methodology, J.W. and L.Y.; software, J.W.; validation, J.W. and Y.Y.; formal analysis, Y.L. and L.Y.; investigation, J.W.; resources, J.W.; data curation, Y.L. and J.W.; writing—original draft preparation, J.W.; writing—review and editing, J.W., Y.L., L.Y. and Y.Y.; visualization, J.W. and Y.Z.; supervision, J.W.; project administration, J.W. and Y.Y.; funding acquisition, J.W. All authors have read and agreed to the published version of the manuscript.

Funding: This work was financially supported by the Scientific Research Foundation of Hunan Provincial Education Department (No. 20B527).

Institutional Review Board Statement: Not applicable.

Informed Consent Statement: Not applicable.

Data Availability Statement: The data presented in this study are available on request from the corresponding author (J.W.), upon reasonable request.

Conflicts of Interest: The authors declare no conflict of interest.

References

1. Wang, J.; Yu, Y.; Dong, Y. Disinfection of Ready-to-Eat Lettuce Using Polyhexamethylene Guanidine Hydrochloride. *Microorganisms* **2020**, *8*, 272. [CrossRef] [PubMed]
2. Wang, J.; Tao, D.; Wang, S.; Li, C.; Li, Y.; Zheng, F.; Wu, Z. Disinfection of lettuce using organic acids: An ecological analysis using 16S rRNA sequencing. *RSC Adv.* **2019**, *9*, 17514–17520. [CrossRef]
3. Wang, J.; Yu, Y.; Dong, Y. Combination of polyhexamethylene guanidine hydrochloride and potassium peroxymonosulfate to disinfect ready-to-eat lettuce. *RSC Adv.* **2020**, *10*, 40316–40320. [CrossRef]

4. Callejón, R.M.; Rodríguez-Naranjo, M.I.; Ubeda, C.; Ortega, R.H.; Garcia-Parrilla, M.C.; Troncoso, A.M. Reported Foodborne Outbreaks Due to Fresh Produce in the United States and European Union: Trends and Causes. *Foodborne Pathog. Dis.* **2015**, *12*, 32–38. [CrossRef]
5. Reports of Selected Escherichia coli Outbreak Investigations. Available online: https://www.cdc.gov/ecoli/outbreaks.html (accessed on 12 June 2021).
6. De Corato, U. Improving the shelf-life and quality of fresh and minimally-processed fruits and vegetables for a modern food industry: A comprehensive critical review from the traditional technologies into the most promising advancements. *Crit. Rev. Food Sci. Nutr.* **2019**, *60*, 940–975. [CrossRef]
7. Ma, L.; Zhang, M.; Bhandari, B.; Gao, Z. Recent developments in novel shelf life extension technologies of fresh-cut fruits and vegetables. *Trends Food Sci. Technol.* **2017**, *64*, 23–38. [CrossRef]
8. Ölmez, H.; Kretzschmar, U. Potential alternative disinfection methods for organic fresh-cut industry for minimizing water consumption and environmental impact. *LWT Food Sci. Technol.* **2009**, *42*, 686–693. [CrossRef]
9. Gil, M.I.; Selma, M.V.; López-Gálvez, F.; Allende, A. Fresh-cut product sanitation and wash water disinfection: Problems and solutions. *Int. J. Food Microbiol.* **2009**, *134*, 37–45. [CrossRef]
10. Luo, Y.; Zhou, B.; Van Haute, S.; Nou, X.; Zhang, B.; Teng, Z.; Turner, E.R.; Wang, Q.; Millner, P.D. Association between bacterial survival and free chlorine concentration during commercial fresh-cut produce wash operation. *Food Microbiol.* **2018**, *70*, 120–128. [CrossRef]
11. Garrido, Y.; Marín, A.; Tudela, J.A.; Allende, A.; Gil, M.I. Chlorate uptake during washing is influenced by product type and cut piece size, as well as washing time and wash water content. *Postharvest Biol. Technol.* **2019**, *151*, 45–52. [CrossRef]
12. Van Haute, S.; Tryland, I.; Escudero, C.; Vanneste, M.; Sampers, I. Chlorine dioxide as water disinfectant during fresh-cut iceberg lettuce washing: Disinfectant demand, disinfection efficiency, and chlorite formation. *LWT* **2017**, *75*, 301–304. [CrossRef]
13. Van Haute, S.; Tryland, I.; Veys, A.; Sampers, I. Wash water disinfection of a full-scale leafy vegetables washing process with hydrogen peroxide and the use of a commercial metal ion mixture to improve disinfection efficiency. *Food Control.* **2015**, *50*, 173–183. [CrossRef]
14. Wang, J.; Zhang, Y.; Yu, Y.; Wu, Z.; Wang, H. Combination of ozone and ultrasonic-assisted aerosolization sanitizer as a sanitizing process to disinfect fresh-cut lettuce. *Ultrason. Sonochem.* **2021**, *76*, 105622. [CrossRef] [PubMed]
15. Zhang, J.; Yang, H. Effects of potential organic compatible sanitisers on organic and conventional fresh-cut lettuce (*Lactuca sativa* Var. Crispa L). *Food Control* **2017**, *72*, 20–26. [CrossRef]
16. Velázquez, L.D.C.; Barbini, N.B.; Escudero, M.E.; Estrada, C.S.M.L.; de Guzmán, A.M.S. Evaluation of chlorine, benzalkonium chloride and lactic acid as sanitizers for reducing Escherichia coli O157:H7 and Yersinia enterocolitica on fresh vegetables. *Food Control* **2009**, *20*, 262–268. [CrossRef]
17. Ölmez, H.; Temur, S. Effects of different sanitizing treatments on biofilms and attachment of Escherichia coli and Listeria monocytogenes on green leaf lettuce. *LWT* **2010**, *43*, 964–970. [CrossRef]
18. Neal, J.A.; Marquez-Gonzalez, M.; Cabrera-Diaz, E.; Lucia, L.M.; O'Bryan, C.A.; Crandall, P.G.; Ricke, S.C.; Castillo, A. Comparison of multiple chemical sanitizers for reducing Salmonella and Escherichia coli O157:H7 on spinach (*Spinacia oleracea*) leaves. *Food Res. Int.* **2012**, *45*, 1123–1128. [CrossRef]
19. Wang, J.; Wang, S.; Sun, Y.; Li, C.; Li, Y.; Zhang, Q.; Wu, Z. Reduction of *Escherichia coli* O157:H7 and naturally present microbes on fresh-cut lettuce using lactic acid and aqueous ozone. *RSC Adv.* **2019**, *9*, 22636–22643. [CrossRef]
20. Ricke, S. Perspectives on the use of organic acids and short chain fatty acids as antimicrobials. *Poult. Sci.* **2003**, *82*, 632–639. [CrossRef]
21. Hui, W.; Colin, O.; Yang, X. Development of a real-time PCR procedure for quantification of viable *Escherichia coli* in popu-lations of *E. coli* exposed to lactic acid, and the acid tolerance of verotoxigenic *E. coli* (VTEC) from cattle hides. *Food Control* **2014**, *43*, 104–109.
22. Nakamura, K.; Tanaka, T.; Kuwahara, A.; Takeo, K. Microassay for proteins on nitrocellulose filter using protein dye-staining procedure. *Anal. Biochem.* **1985**, *148*, 311–319. [CrossRef]
23. Wang, Y.; Qin, Y.; Zhang, Y.; Wu, R.; Li, P. Antibacterial mechanism of plantaricin LPL-1, a novel class IIa bacteriocin against Listeria monocytogenes. *Food Control* **2019**, *97*, 87–93. [CrossRef]
24. Wiśniewski, J.R.; Zougman, A.; Nagaraj, N.; Mann, M. Universal sample preparation method for proteome analysis. *Nat. Methods* **2009**, *6*, 359–362. [CrossRef] [PubMed]
25. Wang, J.; Cheng, Y.; Wu, R.; Jiang, D.; Bai, B.; Tan, D.; Yan, T.; Sun, X.; Zhang, Q.; Wu, Z. Antibacterial Activity of Juglone against Staphylococcus aureus: From Apparent to Proteomic. *Int. J. Mol. Sci.* **2016**, *17*, 965. [CrossRef] [PubMed]
26. Conesa, A.; Gotz, S.; Garciagomez, J.M.; Terol, J.; Talon, M.; Robles, M.J.B. Blast2GO: A universal tool for annotation, visual-ization and analysis in functional genomics research. *Bioinformatics* **2005**, *21*, 3674–3676. [CrossRef]
27. Von Mering, C.; Jensen, L.J.; Snel, B.; Hooper, S.D.; Krupp, M.; Foglierini, M.; Jouffre, N.; Huynen, M.A.; Bork, P. STRING: Known and predicted protein-protein associations, integrated and transferred across organisms. *Nucleic Acids Res.* **2005**, *33*, D433–D437. [CrossRef]
28. Ritter, A.C.; Santi, L.; Vannini, L.; Beys-Da-Silva, W.; Gozzi, G.; Yates, J.; Ragni, L.; Brandelli, A. Comparative proteomic analysis of foodborne Salmonella Enteritidis SE86 subjected to cold plasma treatment. *Food Microbiol.* **2018**, *76*, 310–318. [CrossRef]

29. Ahmed, S.; Zaman, S.; Ahmed, R.; Uddin, N.; Acedo, A.; Bari, L. Effectiveness of non-chlorine sanitizers in improving the safety and quality of fresh betel leaf. *LWT* **2017**, *78*, 77–81. [CrossRef]
30. Finten, G.; Agüero, M.V.; Jagus, R.J. Citric acid as alternative to sodium hypochlorite for washing and disinfection of experimentally-infected spinach leaves. *LWT Food Sci. Technol.* **2017**, *82*, 318–325. [CrossRef]
31. Chen, L.; Zhang, H.; Liu, Q.; Pang, X.; Zhao, X.; Yang, H. Sanitising efficacy of lactic acid combined with low-concentration sodium hypochlorite on Listeria innocua in organic broccoli sprouts. *Int. J. Food Microbiol.* **2019**, *295*, 41–48. [CrossRef]
32. Fan, X.; Sokorai, K.J.; Niemira, B.A.; Mills, R.S.; Zhen, M.Y. Quality of Gamma Ray-irradiated Iceberg Lettuce and Treatments to Minimize Irradiation-induced Disorders. *HortScience* **2012**, *47*, 1108–1112. [CrossRef]
33. Salgado, S.P.; Pearlstein, A.J.; Luo, Y.; Feng, H. Quality of Iceberg (*Lactuca sativa* L.) and Romaine (*L. sativa* L. var. longifolial) lettuce treated by combinations of sanitizer, surfactant, and ultrasound. *LWT* **2014**, *56*, 261–268. [CrossRef]
34. Wang, J.; Sun, Y.; Tao, D.; Li, C.; Zheng, F.; Wu, Z. Reduction of Escherichia coli O157:H7, Listeria monocytogenes, and Naturally Present Microbe Counts on Lettuce using an Acid Mixture of Acetic and Lactic Acid. *Microorganisms* **2019**, *7*, 373. [CrossRef] [PubMed]
35. Vijayakumar, C.; Wolf-Hall, C.E. Evaluation of Household Sanitizers for Reducing Levels of Escherichia coli on Iceberg Lettuce. *J. Food Prot.* **2002**, *65*, 1646–1650. [CrossRef] [PubMed]
36. Poimenidou, S.V.; Bikouli, V.C.; Gardeli, C.; Mitsi, C.; Tarantilis, P.A.; Nychas, G.J.; Skandamis, P.N. Effect of single or com-bined chemical and natural antimicrobial interventions on *Escherichia coli* O157:H7, total microbiota and color of packaged spinach and lettuce. *Int. J. Food Microbiol.* **2016**, *220*, 6–18. [CrossRef]
37. Wang, J.; Liu, D.; Sun, X.; Bai, B.; Jiang, D.; Wu, Z. Label-free quantitative proteomic analysis of the inhibitory activities of juglone against translation and energy metabolism in *Escherichia coli. Phytochem. Lett.* **2016**, *18*, 55–58. [CrossRef]
38. Yang, X.Y.; Zhang, L.; Liu, J.; Li, N.; Yu, G.; Cao, K.; Han, J.; Zeng, G.; Pan, Y.; Sun, X.; et al. Proteomic analysis on the anti-bacterial activity of a Ru(II) complex against Streptococcus pneumoniae. *J. Proteomics* **2015**, *115*, 107–116. [CrossRef]
39. Yang, X.-Y.; Xu, J.-Y.; Meng, M.; Li, N.; Liu, C.-Y.; He, Q.-Y. Dirhodium (II) complex interferes with iron-transport system to exert antibacterial action against Streptococcus pneumoniae. *J. Proteom.* **2018**, *194*, 160–167. [CrossRef]
40. Solis, N.; Parker, B.L.; Kwong, S.M.; Robinson, G.; Firth, N.; Cordwell, S.J. Staphylococcus aureus Surface Proteins Involved in Adaptation to Oxacillin Identified Using a Novel Cell Shaving Approach. *J. Proteome Res.* **2014**, *13*, 2954–2972. [CrossRef]
41. Natale, P.; Brüser, T.; Driessen, A.J.M. Sec- and Tat-mediated protein secretion across the bacterial cytoplasmic membrane—Distinct translocases and mechanisms. *Biochim. Biophys. Acta* **2008**, *1778*, 1735–1756. [CrossRef]
42. Luirink, J.; Hagen-Jongman, C.T.; van der Weijden, C.; Oudega, B.; High, S.; Dobberstein, B.; Kusters, R. An alternative protein targeting pathway in *Escherichia coli*: Studies on the role of FtsY. *EMBO J.* **1994**, *13*, 2289–2296. [CrossRef]
43. Diao, M.; Qi, D.; Xu, M.; Lu, Z.; Lv, F.; Bie, X.; Zhang, C.; Zhao, H. Antibacterial activity and mechanism of monolauroyl-galactosylglycerol against Bacillus cereus. *Food Control* **2018**, *85*, 339–344. [CrossRef]
44. Kang, S.; Kong, F.; Shi, X.; Han, H.; Li, M.; Guan, B.; Yang, M.; Cao, X.; Tao, D.; Zheng, Y.; et al. Antibacterial activity and mechanism of lactobionic acid against Pseudomonas fluorescens and methicillin-resistant Staphylococcus aureus and its application on whole milk. *Food Control* **2020**, *108*, 106876. [CrossRef]
45. Jeckelmann, J.-M.; Harder, D.; Mari, S.A.; Meury, M.; Ucurum, Z.; Muller, D.J.; Erni, B.; Fotiadis, D. Structure and function of the glucose PTS transporter from *Escherichia coli. J. Struct. Biol.* **2011**, *176*, 395–403. [CrossRef]
46. Yamashita, T.; Kawada-Matsuo, M.; Katsumata, T.; Watanabe, A.; Oogai, Y.; Nishitani, Y.; Miyawaki, S.; Komatsuzawa, H. Antibacterial activity of disodium succinoyl glycyrrhetinate, a derivative of glycyrrhetinic acid againstStreptococcus mutans. *Microbiol. Immunol.* **2019**, *63*, 251–260. [CrossRef]
47. Tchieu, J.H.; Norris, V.; Edwards, J.S.; Saier, M.H. The complete phosphotransferase system in Escherichia coli. *J. Mol. Microbiol. Biotechnol.* **2001**, *3*, 329–346.
48. Liu, Y.; Ceruso, M.; Jiang, Y.; Datta, A.R.; Carter, L.; Strain, E.; Pepe, T.; Anastasi, A.; Fratamico, P. Construction ofListeria monocytogenesMutants with In-Frame Deletions in the Phosphotransferase Transport System (PTS) and Analysis of Their Growth under Stress Conditions. *J. Food Sci.* **2013**, *78*, M1392–M1398. [CrossRef]
49. Chakrabarti, A.; Membrez, M.; Morin-Rivron, D.; Siddharth, J.; Chou, C.J.; Henry, H.; Bruce, S.; Metairon, S.; Raymond, F.; Betrisey, B.; et al. Transcriptomics-driven lipidomics (TDL) identifies the microbiome-regulated targets of ileal lipid metabolism. *Npj Syst. Biol. Appl.* **2017**, *3*, 33. [CrossRef] [PubMed]
50. Benga, G. On the definition, nomenclature and classification of water channel proteins (aquaporins and relatives). *Mol. Asp. Med.* **2012**, *33*, 514–517. [CrossRef] [PubMed]
51. Xie, J.; Zhang, A.-H.; Sun, H.; Yan, G.-L.; Wang, X.-J. Recent advances and effective strategies in the discovery and applications of natural products. *RSC Adv.* **2018**, *8*, 812–824. [CrossRef]
52. Zhou, P.; Chen, Y.; Lu, Q.; Qin, H.; Ou, H.; He, B.; Ye, J. Cellular metabolism network of Bacillus thuringiensis related to erythromycin stress and degradation. *Ecotoxicol. Environ. Saf.* **2018**, *160*, 328–341. [CrossRef]
53. Sassanfar, M.; Roberts, J.W. Nature of the SOS-inducing signal in Escherichia coli: The involvement of DNA replication. *J. Mol. Biol.* **1990**, *212*, 79–96. [CrossRef]
54. Kim, E.A.; Panushka, J.; Meyer, T.; Carlisle, R.; Baker, S.; Ide, N.; Lynch, M.; Crane, B.R.; Blair, D.F. Architecture of the Flagellar Switch Complex of *Escherichia coli*: Conformational Plasticity of FliG and Implications for Adaptive Remodeling. *J. Mol. Biol.* **2017**, *429*, 1305–1320. [CrossRef] [PubMed]

55. Ford, K.M.; Antani, J.; Nagarajan, A.; Johnson, M.M.; Lele, P.P. Switching and Torque Generation in Swarming *E. coli*. *Front. Microbiol.* **2018**, *9*, 2197. [CrossRef] [PubMed]
56. Guo, J.; Hu, X.; Gao, Z.; Li, G.; Fu, F.; Shang, X.; Liang, Z.; Shan, Y. Global transcriptomic response of Listeria monocytogenes exposed to Fingered Citron (*Citrus medica* L. var. sarcodactylis Swingle) essential oil. *Food Res. Int.* **2021**, *143*, 110274. [CrossRef] [PubMed]

foods

MDPI

Article

Effect of Cold Shock Pretreatment Combined with Perforation-Mediated Passive Modified Atmosphere Packaging on Storage Quality of Cucumbers

Fucheng Wang [1], Si Mi [1], Bimal Chitrakar [1], Jinsong Li [2] and Xianghong Wang [1,*]

[1] College of Food Science and Technology, Hebei Agricultural University, Baoding 071001, China; laster2022@163.com (F.W.); misi@hebau.edu.cn (S.M.); bimal@hebau.edu.cn (B.C.)

[2] Bureau of Agriculture and Rural Areas of Qinhuangdao, Qinhuangdao 066000, China; april202204@163.com

* Correspondence: wangxianghong@hebau.edu.cn; Tel.: +86-(0312)-7528195

Abstract: This study evaluated the application of cold shock combined with perforation-mediated passive modified atmosphere packaging technology (CS-PMAP) for cucumber preservation through physicochemical, sensory, and nutritional qualities. The effectiveness of CS-PMAP in maintaining the quality of fresh cucumbers was studied; cucumbers were pretreated with cold shock and then packed into perforated polyethylene bags (bag size of 20 × 30 cm; film thickness of 0.07 mm; and two holes in each bag with a diameter of 6 mm), while the cucumbers without cold shock were considered as the control. Storage of the samples was performed at (13 ± 2) °C for 20 days to determine the quality changes in terms of gas composition, weight loss, skin color, texture, total soluble solids (TSS), ascorbic acid, malondialdehyde (MDA), and volatile organic compounds (VOCs). The CS-PMAP showed a significant improvement in maintaining firmness, TSS, ascorbic acid, and flavor profile of cucumbers; the control samples without cold shock showed higher weight loss and MDA levels. Results of this study confirmed that CS-PMAP has potential use in the storage of cucumbers.

Keywords: cold shock; MAP; cucumber; storage; VOCs

Citation: Wang, F.; Mi, S.; Chitrakar, B.; Li, J.; Wang, X. Effect of Cold Shock Pretreatment Combined with Perforation-Mediated Passive Modified Atmosphere Packaging on Storage Quality of Cucumbers. *Foods* **2022**, *11*, 1267. https://doi.org/10.3390/foods11091267

Academic Editor: Peng Jin

Received: 5 April 2022
Accepted: 26 April 2022
Published: 27 April 2022

1. Introduction

Cucumbers (*Cucumis sativus* L.) are cylindrical fruits, normally considered as vegetables because of the way they are used in food preparation; they are widely cultivated in China [1]. They are rich in minerals, vitamins, and flavonoids but also high in moisture, which makes them susceptible to deterioration after harvest due to water loss, yellowing color, and microbial action; the shelf life of fresh cucumbers is limited to 3–5 days at room temperature [2]. To maintain the post-harvest quality of cucumbers, several treatment methods have been studied, including chemical preservatives [3], physical treatments [4,5], coatings [2,6,7], modified atmosphere packaging [8], and co-preservations [9,10]. However, with a rising demand of safe foods from consumers, a safer and more effective method of preservation to replace the traditional preservation methods for cucumber is very important.

The pre-cooling process is crucial in the cold supply chain management of vegetables and fruits. In recent years, cold shock treatment (CS) has received extensive attention as a modified pre-cooling method. It is a physical preservation method of fruits and vegetables, which is usually performed with ice water [11]. Simple, convenient, safe, and low cost are the characteristics of CS treatment. Therefore, it is suitable for high-volume processing in both small and large scales industries, especially in developing countries. Several studies have shown that cold shock is able to maintain the quality of sweet cherry [11]; reduce lignification of asparagus [10]; inhibit the decrease of firmness in banana [12]; and extend the shelf life of avocado fruits [13].

Modified atmosphere packaging (MAP) is a method to slow down food deterioration by using different kinds of packaging films and different concentrations of oxygen (O_2) and carbon dioxide (CO_2). MAP can be done in two ways: The first one is active MAP, where a product-specific gas or gas mixture is used in the pack, while the second one is passive MAP, where a spontaneous change in the gas composition occurs from the fresh food that respires in the packaged MAP [14] and no additional gas supply is practiced. Therefore, passive MAP is more cost-effective than active MAP and is reported to be applicable to a wide range of fresh fruits and vegetables [15]. Furthermore, for passive MAP, the choice of packaging films is more important. Agricultural products consume O_2 through respiration and produce CO_2 and water vapor during storage [16]. Most films commonly used in MAP have limitations in permeation properties, including high barrier properties to water vapor and unbalanced gas transmission rate, resulting in a favorable condition for anaerobic bacteria and the development of strange odors, among others. It has been reported that O_2 and CO_2 levels obtained with conventional MAP are rarely sufficient to maintain high respiratory product quality throughout storage [17]. Perforations with suitable numbers and dimensions have been discovered to promote the packaging film permeability and quickly achieve an O_2 and CO_2 balance inside the package [18].

The present work was aimed to evaluate the application of perforation-mediated passive modified atmosphere packaging (PM-PMAP) treatment, combined with CS for maintaining the post-harvest quality of cucumber by assessing the physicochemical, sensory, and nutritional properties of cucumber.

2. Materials and Methods

2.1. Cucumber Pretreatments and Packaging

Cucumbers were harvested from a vegetable farm located in Changli County, Qinhuangdao City, Hebei Province, China (36°57′ N, 119°45′ E). The picked cucumbers were immediately transported to the laboratory and were sorted based on maturity (color, gloss, and shape) [19], soundness (mechanical injury and insect infestation), and size. Polyethylene (PE) film is colorless, odorless, non-toxic, tasteless, high in mechanical strength, and has moisture resistance in humid environments. Studies have shown that PE has a good preservation effect on fresh fruits and vegetables such as cucumbers, persimmons, and mushrooms [15], thus making it the first choice as packaging material for these products. The sorted cucumbers were randomly divided into four groups, each group containing 45 cucumbers. The first group was immersed in 0 °C ice water for 20 min (CS20). Similarly, the second and third groups were soaked in ice water at 0 °C for 40 min (CS40) and 60 min (CS60), respectively, while the fourth group was not pretreated with ice water (NCS). Then, all groups of cucumbers were put into perforated PE film bags (bag size was 20 × 30 cm; film thickness was 0.07 mm; each bag had two holes having a diameter of 6 mm) and then heat-sealed. Groups were defined as CS20+P, CS40+P, CS60+P, and NCS+P (control group). Finally, all the samples were stored at 13 ± 2 °C for 20 days in a cold storage room; the reason behind choosing that temperature was that cucumbers are susceptible to chill injury at 10 °C or lower temperatures, while they turn yellow at 15 °C or higher [8,20]. Samplings were done on the 0, 4, 8, 12, 16, and 20 days to analyze the following quality attributes: internal atmosphere composition, weight loss, instrumental color, instrumental texture, total soluble solids, ascorbic acid content, malondialdehyde (MDA) content, and volatile organic compounds (VOCs).

2.2. Internal Atmosphere Composition

O_2 and CO_2 concentrations inside the packages were measured using a portable gas analyzer (OXYBABY® 6.0, Witt, Witten, Germany). The instrument measured O_2 and CO_2 concentrations in the headspace through electrochemical and infrared sensors. Results were expressed as percentage (%) of O_2 and CO_2 inside the bags. Triplicate measurements were done for each group.

2.3. Weight Loss

An electronic scale with an accuracy of 0.1 g was used to weigh the sample at each sampling time, and the weight loss (%) was calculated based on the initial weight. All measurements were done in triplicate.

2.4. Skin Color

The color of cucumbers was measured with a chromameter (CR-400, Konica Minolta, Tokyo, Japan). The parameters L^* (lightness), a^* (red and green chromaticity), and b^* (blue and yellow chromaticity) values were used to describe the colors. The instrument was calibrated using a standard white reflector before measurement. For each experimental group, five cucumbers were randomly selected, and triplicate measurements were made for each sample.

2.5. Instrumental Texture

Texture was determined according to Yang et al. [21] with slight modifications. A texture analyzer (TMS-Pro, Food Technology Corporation, Sterling, VA, USA) was used to conduct texture profile analysis using 1 cubic cm cucumber samples. A cylindrical probe with a diameter of 35 mm was used at the probe height of 20 mm. A test speed of 1 mm/s was used throughout (before, during, and after) the test with a depression distance of 50% of the compression degree, while the contact force used was 0.5 N. Firmness was expressed as maximum compression force in Newtons (N) and five measurements were taken for each experimental group.

2.6. Total Soluble Solids

Cucumbers were mashed into a fine paste using a lab blender. The mash was then squeezed through four layers of cheesecloth, and the juice was used for the TSS assay using a portable digital refractometer (RSD200, AS ONE, Osaka, Japan). Measurements were made on five fruits per experimental group.

2.7. Ascorbic Acid

Ascorbic acid content was determined according to Yang et al. [22] with slight modifications. Accurately weighed (20 g) cucumber was put into a crusher to mash into a homogenate with 20 g of the oxalic acid solution The homogenized sample was made to 100 mL with oxalic acid solution in a volumetric flask. The filtered aliquot (10 mL) was titrated with standard 2,6-dichloroindophenol solution until pink color that last for 15 s. The whole process was carried out under dark conditions. The content of ascorbic acid in cucumber samples was calculated as follows.

$$X\ (\text{mg per 100 g}) = \frac{(V - V_0) \times T \times A}{m} \times 100 \quad (1)$$

where X is the content of ascorbic acid in milligrams per hundred grams (mg/100 g); V is the volume of 2,6-dichloroindophenol solution consumed for titration (mL); V_0 is the volume of 2,6-dichloroindophenol solution consumed for titration blank (mL); T is milligrams of ascorbic acid per milliliter of 2,6-dichloroindophenol solution (mg/mL); A is dilution factor; and m is the mass of the sample (g).

2.8. Malondialdehyde

Malondialdehyde (MDA) content was measured using the method described by Wang et al. [23] with slight modifications. MDA content was determined using a microplate spectrophotometer (Multiskan Spectrum, Thermo Scientific, Waltham, MA, USA). Cucumber sample (2 g) was ground with 5 mL of trichloroacetic acid (TCA; 10%, *w/v*). After centrifugation at 8000 r/min for 10 min, 2 mL supernatant (2 mL TCA for control) was mixed with 2 mL of thiobarbituric acid (TBA; 0.67%, *w/v*). The mixture was put in a boiling water bath for 20 min. The sample was cooled rapidly and again centrifuged at 8000 r/min for 3 min. The supernatant was taken to determine the optical density (OD) at 532 nm, 600 nm, and 450 nm. Five replicates were performed for each treatment. The MDA content was calculated according to following formula:

$$\text{MDA content}\left(\frac{\mu\text{mol}}{\text{gm}_\text{F}}\right) = \frac{[6.45 \times (OD_{532} - OD_{600}) - 0.56 \times OD_{450}] \times V_1 \times V_2}{V_3 \times m \times 1000} \quad (2)$$

where 6.45 is the absorbance correction coefficient at wavelength 532 nm; 0.56 is the absorbance correction coefficient at wavelength 450 nm; OD_{532}, OD_{600}, and OD_{450} are the absorbance values at 532, 600, and 450 nm, respectively; V_1 is the total volume of sample extract (mL); V_2 is the total volume of reaction solution (mL); V_3 is the volume of sample extracts taken (mL); and m is the mass of the sample taken (g).

2.9. Volatile Organic Compounds

The volatile organic compounds were analyzed following the methods of Li et al. [24], and Song et al. [25] with some modification using gas chromatography–ion mobility spectrometry (GC–IMS) (FlavourSpec®, Gesellschaft für Analytische Sensorsysteme mbH, Dortmund, Germany), equipped with an autosampler unit (CTC Analytics AG, Zwingen, Switzerland) that can directly sample from the headspace by using a 1 mL air-tight heated syringe.

Cucumber samples (2.5 g) were cut into small pieces and placed into 20 mL headspace glass sampling vials. Subsequently, samples were incubated at 40 °C for 10 min with an oscillator speed of 500 r/min. After incubation, 1 mL of headspace gas was automatically injected through a syringe at 85 °C. Then, the samples were driven into a GC column by nitrogen at a programmed flow as follows: 2 mL/min for 2 min; linearly increasing to 15 mL/min from 2 to 10 min; linearly increasing to 100 mL/min from 10 to 25 min; held for 5 min. The column type was MTX-5 (15 m × 0.53 mm, 1 µm). The syringe was automatically flushed with a stream of nitrogen before and after each analysis to avoid cross-contamination.

The retention index (RI) of each compound was calculated using n-ketones C4–C9 as external references. Identification of volatile compounds was based on comparison of RI and drift times (Dts) with GC–IMS libraries.

2.10. Statistical Analysis

Microsoft Office Excel 2016 was used to perform the basic processing of the obtained experimental data; other statistical analyses were performed using Addinsoft (2021) (XLSTAT statistical and data analysis solution, New York, NY, USA). A basic descriptive statistical analysis was followed by an analysis of variance test (ANOVA) for mean comparisons; 95% confidence intervals ($p < 0.05$) were set throughout the data analysis to identify significant differences. Principal component analysis and partial least squares discrimination analysis were also performed. VOCs were collected and analyzed using the VOCal analysis software accompanying the GC–IMS instrument. The differences in fingerprint profiles were compared by the reporter and galerie plug-ins equipped with the GC–IMS instrument. The VOCs were qualitatively analyzed by applying the NIST and IMS databases built into the Library Search software.

3. Results and Discussion

3.1. Internal Atmosphere Composition

It is well known that the changes in gas composition in a MAP system are mainly dependent on storage temperature, film permeability, and product respiration rate [26]. As expected, a decrease in O_2 concentration as well as an increase in CO_2 concentration were observed during the storage period at 13 °C (Figure 1A,B). All the treatment groups showed a rapid decrease in O_2 concentrations during days 4 to 12 ($p < 0.05$); the decrease was slowed down during the subsequent storage period. On day 20, the O_2 concentrations in the CS20+P, CS40+P, CS60+P, and NCS+P groups decreased from 23.01% to 21.84%,

21.48%, 21.40%, and 21.37%, respectively. The O_2 in the CS40+P group declined the slowest. Moreover, there was a significant difference between the NCS+P group and the CS40+P group ($p < 0.05$) from the fourth day onwards. When cucumbers were treated by cold shock, the temperature of cucumbers lowered rapidly. The property and structure of the protease related to respiration may be influenced by the cold shock, which can reduce the respiration, so the O_2 concentration decreased slowly. All the treatment groups showed an increase in CO_2 levels during the storage period; CS20+P, CS40+P CS60+P, and NCS+P groups exhibited an increase from 0 to 2.40%, 1.87%, 2.57%, and 3.23%, respectively, on day 20. The CO_2 concentration in the CS40+P group increased the slowest. In addition, the CO_2 concentration in the CS40+P group was significantly lower than the other two CS treatment groups from the fourth day. It may be that the treatment time of CS20+P was too short, resulting in insufficient reduction of enzyme activities related to respiration, and that the treatment time of CS60+P was too long, causing some irreversible damage to cucumber tissue. In summary, the experimental results indicated that CS regulated effectively the concentrations of O_2 and CO_2 in the packaging bags. Similar results were observed in avocado fruits subjected to cold shock with 0 °C ice-water mixture [13]. The CS40+P treatment best inhibited the respiration rate of cucumber fruits, which was beneficial to cucumber storage.

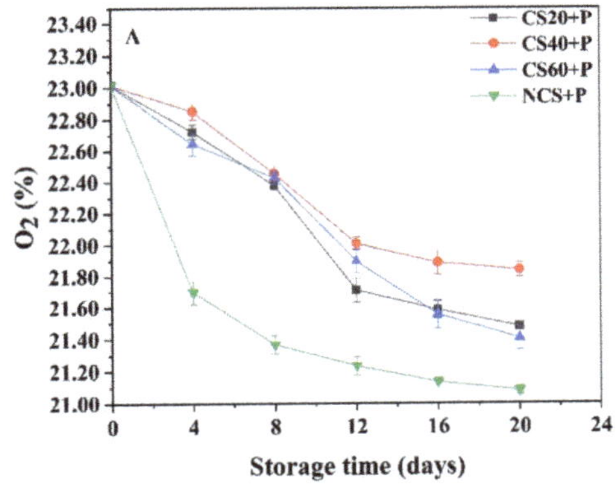

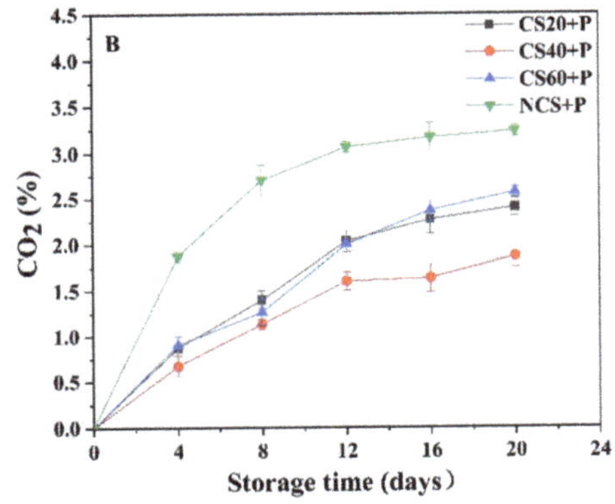

Figure 1. Changes in O_2 (**A**) and CO_2 (**B**) content during storage at 13 ± 2 °C in different treatments. CS20+P—packaged after 20 min of cold shock; CS40+P—packaged after 40 min of cold shock; CS60+P—packaged after 60 min of cold shock; NCS+P—packaged after 0 min of cold shock.

3.2. Weight Loss

Cucumbers have a high moisture content (about 95%) [8]. One of the main problems during post-harvest storage of cucumbers is rapid water loss from cucumbers, resulting in weight loss. This may be because cucumbers are only protected by a thin skin structure, and the weight loss of fresh produce in the postharvest process is caused by two factors, namely transpiration and respiration [27]. Transpiration can cause water loss, while respiration can cause dry matter loss [28]. Harvested produce releases water vapor into the surrounding atmosphere through transpiration, while the respiration process uses reserves of organic materials and also releases water vapor [29] to reduce the weight of fruits and vegetables.

Weight loss of cucumbers increased consistently with the prolongation of storage time at 13 °C (Table 1). Started from day 12, weight loss of the NCS+P group was significantly ($p < 0.05$) higher than the other groups. In general, CS-treated samples showed less weight loss compared to NCS-treated samples, and the lowest weight loss was observed in the CS40+P group. This results revealed that CS-treated cucumbers could retain their weight, compared to those with NCS, and the CS40+P treatment was the best combination to prevent cucumber weight loss. The above results showed that cold shock may inhibit the respiration and transpiration of cucumber. This was in agreement with the previous studies on asparagus spears, where the reductions in fresh weight loss by CS were significant during the storage period [10].

Table 1. Effect of different treatments on weight loss of cucumbers during storage at 13 ± 2 °C. CS20+P—packaged after 20 min of cold shock; CS40+P—packaged after 40 min of cold shock; CS60+P—packaged after 60 min of cold shock; NCS+P—packaged after 0 min of cold shock.

Storage Time (days)	Weight Loss (%)			
	CS20+P	CS40+P	CS60+P	NCS+P
0	0.00 ± 0.00 j	0.00 ± 0.00 j	0.00 ± 0.00 j	0.00 ± 0.00 j
4	0.14 ± 0.05 h,i	0.08 ± 0.01 i,j	0.15 ± 0.03 h,i	0.12 ± 0.08 h, i,j
8	0.20 ± 0.03 h,i	0.13 ± 0.02 h,i,j	0.25 ± 0.06 h	0.20 ± 0.03 h,i
12	0.74 ± 0.04 f	0.52 ± 0.06 g	0.62 ± 0.07 f,g	0.97 ± 0.14 d,e
16	1.00 ± 0.03 d,e	0.67 ± 0.06 f	0.94 ± 0.09 e	1.51 ± 0.04 c
20	1.90 ± 0.14 b	1.09 ± 0.06 d	1.93 ± 0.06 b	2.32 ± 0.27 a

Values with different letters in each row are significantly different ($p < 0.05$).

3.3. Skin Color

Color is one of the most important quality parameters of consumer acceptance [30], and color changes are important indicators of shelf-life and maturity of cucumbers. The L^* values of all treated cucumbers increased with storage period, which was in agreement with a previous study [8]. The L^* value of the cucumbers in the NCS+P group remained higher than that of the other groups, and after the 8th day of storage, the L^* value of the cucumbers in the NCS+P group was significantly ($p < 0.05$) higher than that of the CS-treated groups. Among the cold shock groups, the L^* value of cucumbers in the CS40+P group had the smallest change, namely from 48.56 to 51.68. The results from Figure 2A show that the brightness of cucumbers increased with the extension of the storage period, which may be because of the effect of light, oxidation, and respiration that caused the cucumber to change from its natural green color to white [8]. The cucumbers in the CS40+P group had the smallest change; it may be that CS40+P treatment inhibited the respiration and oxidation of cucumber.

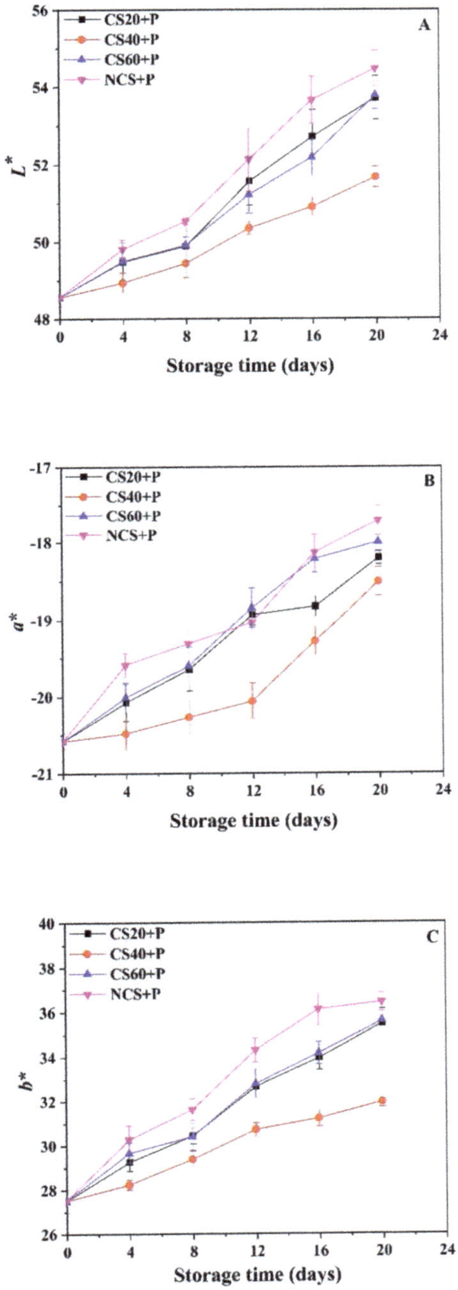

Figure 2. Effect of different treatments on the *L** (**A**), *a** (**B**), and *b** (**C**) values of cucumbers during storage at 13 ± 2 °C. CS20+P—packaged after 20 min of cold shock; CS40+P—packaged after 40 min of cold shock; CS60+P—packaged after 60 min of cold shock; NCS+P—packaged after 0 min of cold shock.

During the storage period, the a^* values of CS20+P, CS40+P, CS60+P, and NCS+P groups increased from −20.58 to −18.20, −18.51, and −17.99, respectively. The a^* values of these three CS groups were significantly ($p < 0.05$) lower than those of the NCS group on day 4 and day 8, and the a^* value of the CS40+P group was consistently maintained significantly ($p < 0.05$) lower than that of other groups. In other words, the greenness of cucumbers in all groups was getting lighter, and the CS40+P group showed the least change. The green color of cucumber is mainly attributed to the presence of chlorophylls, and the decrease in greenness during storage may be caused by the decomposition of chlorophylls. The mechanisms of cold shock treatment are partially based on a decrease in chlorophyllase activity that results in delaying the degradation of chlorophylls [31].

The b^* values of all groups increased significantly ($p < 0.05$) with time during the storage period. The mean values of b^* for CS20+P, CS40+P, CS60+P, and NCS+P groups increased from 27.52 to 35.50, 31.97, 35.62, and 36.44, respectively, indicating that the color of the cucumber was changing towards a yellow color. The yellow color of cucumber may be caused by the accumulation of carotenoids and flavonoids in the rind [32]. The b^* values of the CS-treated group were significantly ($p < 0.05$) lower than those of the NCS-treated group from the eighth day onwards. Moreover, the b^* values of the CS40+P group were significantly ($p < 0.05$) lower than those of the CS20+P, CS60+P, and NCS+P groups from day 4 onwards.

The color (L^*, a^*, and b^* values) of cucumbers in CS20+P, CS60+P, and NCS+P groups changed faster than that of the CS40+P group. It may be that the CS20+P treatment time was too short to improve the stress resistance mechanism of cucumber; the CS60+P treatment time was too long, which accelerated the aging of cucumber; and CS40+P treatment improved the stress resistance of cucumber and delayed the color change. In summary, CS40+P treatment had minimal color change; thus, it had a positive effect on the color protection of postharvest cucumbers.

3.4. Instrumental Texture

Firmness is a very important indicator of the quality and shelf-life of agricultural products, as it reflects the biochemical changes that take place in the cell structure [33]. Cucumbers showed a rapid loss of firmness during storage, which was mainly due to water loss [34]. The firmness of cucumbers in all groups decreased gradually during the storage period (Figure 3). Previous research has shown that the hardness of cucumber decreased with storage time [35] as found in our experimental results. At the end of storage, the hardness of the NCS+P group decreased by 55.2% (from 47.48 N to 21.12 N) and was significantly ($p < 0.05$) lower than that of the CS groups. The CS40+P treatment maintained high hardness values of cucumbers throughout the storage period. It may be that the cold shock time in the CS20+P group was not enough to improve the resistance of cucumbers and reduce water loss, and that the cold shock time in the CS60+P group was too long, which made the cucumbers suffer from cold damage and accelerated softening. CS and MAP have been reported to show beneficial effects in preserving texture [36,37], possibly by attenuating respiration and transpiration of cucumber, thereby reducing firmness loss.

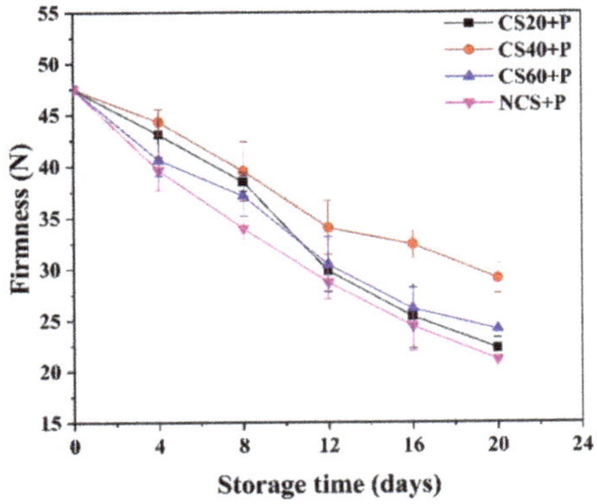

Figure 3. Effect of different treatments on the firmness of cucumbers during storage at 13 ± 2 °C. CS20+P—packaged after 20 min of cold shock; CS40+P—packaged after 40 min of cold shock; CS60+P—packaged after 60 min of cold shock; NCS+P—packaged after 0 min of cold shock.

3.5. Total Soluble Solids (TSS)

The TSS mainly include sugars and acids; TSS is an important parameter to identify fruit maturity and quality [3]. The senescence of fruits is generally involved in a decrease in TSS. The post-harvest deformation of fresh cucumbers (wilting) is caused by changes in water and polysaccharide content that degrade the cell wall [38]. The TSS of all groups gradually reduced with storage time (Figure 4). The decrease of TSS indicated that the cucumber was aging. TSS participated in carbohydrate metabolism of cells [39], thus, causing a reduction of TSS. At end of the 20-day storage period, the TSS value of the CS40+P treatment group was 3.02, which was significantly ($p < 0.05$) higher than that of the other treatment groups. This may be due to the physical effects of CS40+P treatment, which may have inhibited the activity of enzymes associated with carbohydrate metabolism as well as cucumber respiration, thereby maintaining a high TSS level, and the PM-PMAP reduced the gas exchange between the samples and the atmosphere to lower the metabolic rate and slow the hydrolysis of carbohydrates [40].

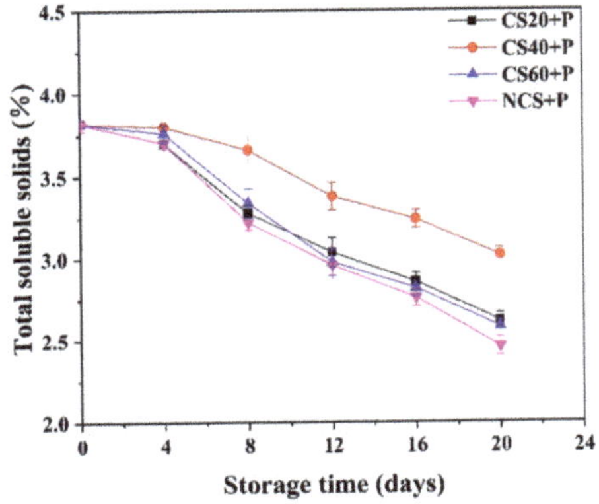

Figure 4. Effect of different treatments on the total soluble solids content of cucumbers during storage at 13 ± 2 °C. CS20+P—packaged after 20 min of cold shock; CS40+P—packaged after 40 min of cold shock; CS60+P—packaged after 60 min of cold shock; NCS+P—packaged after 0 min of cold shock.

3.6. Ascorbic Acid

Ascorbic acid is a key nutritional indicator in fruits and vegetables [41] and is an important ingredient to resist oxidant reactions. As can be seen from Figure 5, the ascorbic acid of all treatment groups decreased with time and was significantly ($p < 0.05$) lower than that of fresh cucumber at day 20. Cucumbers in the NCS+P treatment group showed the greatest decline (from 12.24 to 3.33 mg/100g), and the CS40+P treatment group showed the least decrease (from 12.24 to 4.50 mg/100g). From day 4, the CS40+P group was significantly ($p < 0.05$) higher than other three groups. The reason may be that perforated packaging bags formed a good gas balance between cucumber and the outside world, restricted gas exchange, and reduced the oxidation of ascorbic acid. The beneficial effect of polyethylene-based MAP in retaining ascorbic acid in cucumber, green chilies, and jujube was acknowledged by various studies [9,42,43]. In addition, the combination of cold shock treatment and PM-PMAP reduced the respiration rate of cucumber to reduce the consumption of ascorbic acid, which can be reduced by reducing the respiratory rate of the sample [44].

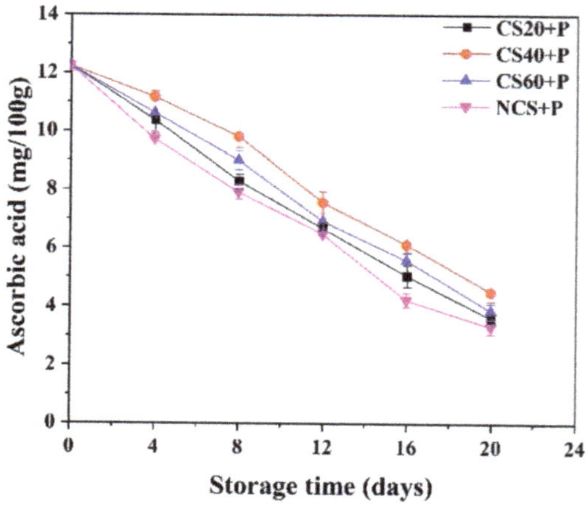

Figure 5. Effect of different treatments on the ascorbic acid content of cucumbers during storage at 13 ± 2 °C. CS20+P—packaged after 20 min of cold shock; CS40+P—packaged after 40 min of cold shock; CS60+P—packaged after 60 min of cold shock; NCS+P—packaged after 0 min of cold shock.

3.7. Malondialdehyde (MDA)

As the cucumber fruit matures, the membrane lipid function weakens due to membrane lipid peroxidation. MDA reflects lipid peroxidation related to the permeability and integrity of the membrane [45]. MDA is an indicator of membrane damage [46], and its content can reflect the stress tolerance of plants [47]. Figure 6 shows that the amount of MDA in all treatment groups increased during the storage period. At the beginning of storage, the MDA content in cucumber samples was 0.0022 μmol/g, while at the end of storage, the MDA contents of CS20+P, CS40+P, CS60+P, and NCS+P groups increased to 0.0046, 0.0036, 0.0046, and 0.0048 μmol/g, respectively. The largest increase in MDA was in the control group. From day 8, the MDA content of the CS40+P group was significantly ($p < 0.05$) lower than the other three groups. It was shown that CS40+P treatment had a positive effect on maintaining the antioxidant capacity of plant lipids. It may be that CS20+P treatment time was too short to improve the antioxidant capacity of cucumbers, and that CS60+P treatment time was too long, causing irreversible damage to cucumber tissue and accelerating cucumber oxidation. The study on sweet cherry showed that cold shock treatment inhibited MDA accumulation [11], which was consistent with our findings.

This may be because CS treatment reduced the enzymatic activity and water mobility to reduce the accumulation of MDA.

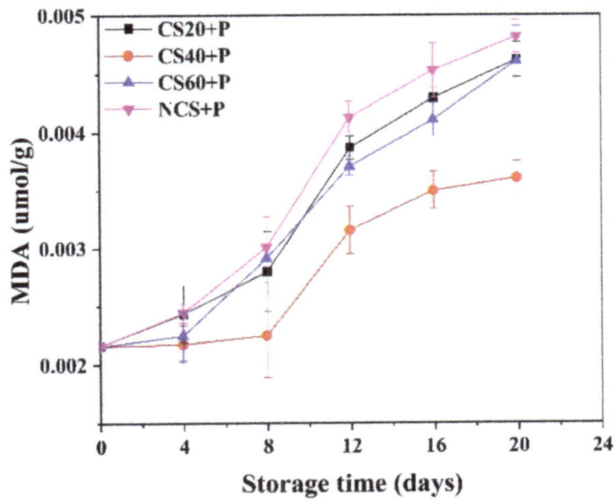

Figure 6. Effect of different treatments on the MDA content of cucumbers during storage at 13 ± 2 °C. CS20+P—packaged after 20 min of cold shock; CS40+P—packaged after 40 min of cold shock; CS60+P—packaged after 60 min of cold shock; NCS+P—packaged after 0 min of cold shock.

3.8. Volatile Organic Compounds

The separation and identification of VOCs in cucumber samples were carried out by the GC–IMS technique. Figure 7 shows two-dimensional difference contrast spectra formed by using the spectrum of cucumber at 0 day as the reference and other spectra deducted from the reference. The horizontal coordinates indicate the ion migration time (drift time, Dt), and the vertical coordinates indicate the retention time (Rt) of gas chromatography. The migration time of VOCs in the samples ranged from 0 to 1.6 ms, and the retention time was concentrated between 100 and 900 s. All VOCs were detected within 30 min. The vertical line at the horizontal coordinate 1.0 in the figure indicates the reactive ion peak (RIP); each dot on both sides of the RIP represent a VOC. In order to distinguish different volatile profiles of CS20+P, CS40+P, CS60+P, and NCS+P, the reactive peak was normalized. In the subtracted spectra, the same concentration of the substance is offset as white; red color indicates that the concentration of the substance is higher than the reference value; the blue color indicates that the concentration of the substance is lower than the reference value; and the darker color indicates a greater difference [48]. Some organic substances appeared at different migration times and formed two or even more signal peaks in the spectrum, due to the higher concentration of the substance; two or more molecules shared a proton or electron, forming a dimer or even a multimer. It can be seen from the figure that the information of gas-phase ion mobility spectra of cucumber samples under different treatment conditions showed some differences during storage, indicating that the content of volatile organic substances in cucumber changed during storage.

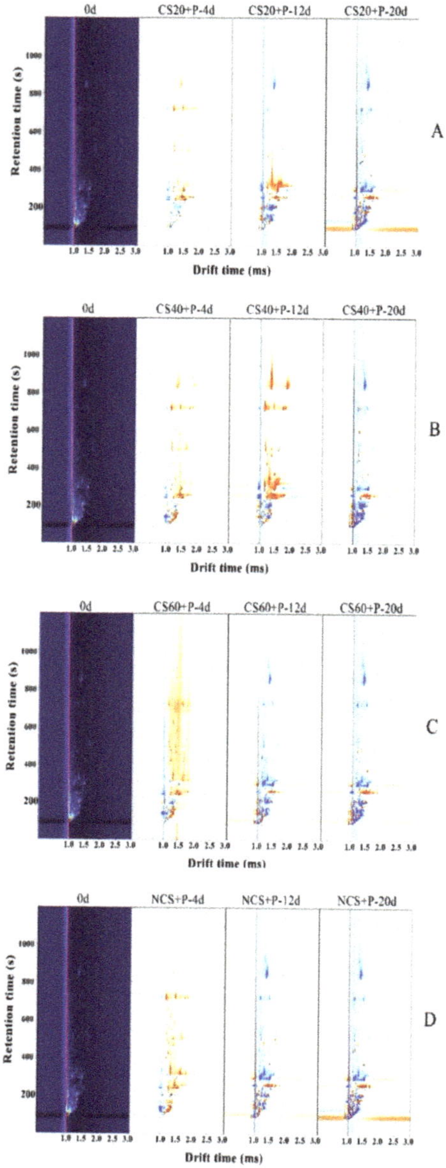

Figure 7. Comparison of the two-dimensional difference spectra of volatile organic compounds GC–IMS in cucumbers under different treatments (**A**): CS20+P; (**B**): CS40+P; (**C**): CS60+P; (**D**): NCS+P.CS20+P—packaged after 20 min of cold shock; CS40+P—packaged after 40 min of cold shock; CS60+P—packaged after 60 min of cold shock; NCS+P—packaged after 0 min of cold shock.

In Table 2, a total of 53 volatile components (monomers and dimers of some substances) could be unambiguously characterized, including 13 alcohols, 18 aldehydes, 6 ketones, 5 esters, 4 acids, 2 ethers, and 5 others.

Table 2. Qualitative information of volatile organic compounds in cucumbers under different treatments CS20+P; CS40+P; CS60+P; and NCS+P. CS20+P—packaged after 20 min of cold shock; CS40+P—packaged after 40 min of cold shock; CS60+P—packaged after 60 min of cold shock; NCS+P—packaged after 0 min of cold shock.

Count	Compounds	Formula	MW	RI	Rt/s	Dt/ms
1	(E,Z)-2,6-nonadienal	$C_9H_{14}O$	138.2	1160.1	841.643	1.37517
2	nonanal	$C_9H_{18}O$	142.2	1106	722.887	1.47299
3	(E, E)-2,4-heptadienal	$C_7H_{10}O$	110.2	1017.3	528.284	1.19128
4	2-ethyl-5-methylpyrazine	$C_7H_{10}N_2$	122.2	1004.3	499.76	1.19805
5	2,4-heptadienal	$C_7H_{10}O$	110.2	990.3	473.833	1.18517
6	hexanoic acid	$C_6H_{12}O_2$	116.2	989.7	472.939	1.30197
7	benzaldehyde	C_7H_6O	106.1	967.1	440.753	1.14867
8	octanal	$C_8H_{16}O$	128.2	1005.4	502.228	1.39972
9	2-ethyl-5-methylpyrazine	$C_7H_{10}N_2$	122.2	1005.4	502.352	1.67368
10	(E)-hept-2-enal	$C_7H_{12}O$	112.2	956	424.963	1.25224
11	(E)-2-hexenal	$C_6H_{10}O$	98.1	851.2	296.851	1.17821
12	(Z)-3-hexen-1-ol	$C_6H_{12}O$	100.2	847.5	293.517	1.51382
13	n-hexanol	$C_6H_{14}O$	102.2	875.5	318.735	1.32264
14	hexanal	$C_6H_{12}O$	100.2	793.2	244.5	1.25877
15	2-methylpropanoic acid	$C_4H_8O_2$	88.1	772.1	228.908	1.15595
16	(E)-2-pentenal	C_5H_8O	84.1	746.8	213.315	1.10294
17	3-hydroxybutan-2-one	$C_4H_8O_2$	88.1	719.7	196.612	1.06159
18	3-hydroxybutan-2-one	$C_4H_8O_2$	88.1	717.1	194.999	1.32669
19	3-pentanone	$C_5H_{10}O$	86.1	688.9	177.662	1.11033
20	3-pentanone	$C_5H_{10}O$	86.1	687.2	176.856	1.35328
21	3-methylbutanal	$C_5H_{10}O$	86.1	642.7	158.107	1.1982
22	1,2-dimethoxyethane	$C_4H_{10}O_2$	90.1	646.7	159.822	1.29793
23	acetic acid ethyl ester	$C_4H_8O_2$	88.1	608.5	143.747	1.09566
24	acetic acid ethyl ester	$C_4H_8O_2$	88.1	605.4	142.442	1.33452
25	1-propanol	C_3H_8O	60.1	570.2	127.59	1.11288
26	butanal	C_4H_8O	72.1	592.3	136.896	1.28754
27	2,3-butanediol	$C_4H_{10}O_2$	90.1	792.5	243.869	1.36382
28	3-methylbutanal	$C_5H_{10}O$	86.1	645.7	159.384	1.39846
29	n-Hexanol	$C_6H_{14}O$	102.2	874.5	317.831	1.63758
30	hexanal	$C_6H_{12}O$	100.2	792.5	243.89	1.55909
31	1-propene-3-methylthio	C_4H_8S	88.2	700.9	185.033	1.04133
32	2-hexanol	$C_6H_{14}O$	102.2	810.1	259.734	1.58033
33	pentan-1-ol	$C_5H_{12}O$	88.1	766.5	225.463	1.24878
34	2-methyl-1-propanol	$C_4H_{10}O$	74.1	625.4	150.847	1.17441
35	benzaldehyde	C_7H_6O	106.1	969	443.513	1.46407
36	2-methylbutanoic acid	$C_5H_{10}O_2$	102.1	876.5	319.684	1.47052
37	heptanal	$C_7H_{14}O$	114.2	901	346.449	1.32688
38	benzene acetaldehyde	C_8H_8O	120.2	1054.9	610.944	1.25142
39	2-pentyl furan	$C_9H_{14}O$	138.2	992.9	477.61	1.25149
40	N,N-diethylethanamine	$C_6H_{15}N$	101.2	688.7	177.505	1.21985
41	2-methylbutan-1-ol	$C_5H_{12}O$	88.1	739.1	208.58	1.46753
42	2,3-butanediol	$C_4H_{10}O_2$	90.1	826.6	274.661	1.36211
43	methyl butyrate	$C_5H_{10}O_2$	102.1	741.9	210.305	1.15562
44	ethyl butanoate	$C_6H_{12}O_2$	116.2	759.3	221.019	1.20389
45	3-butenenitrile	C_4H_5N	67.1	621.5	149.205	1.12862
46	propanoic acid	$C_3H_6O_2$	74.1	742.2	210.487	1.27215
47	hydroxyacetone	$C_3H_6O_2$	74.1	718.6	195.924	1.22934
48	2-propanol	C_3H_8O	60.1	537.8	113.95	1.16127
49	1-hydroxy-2-propanone	$C_3H_6O_2$	74.1	671	170.026	1.04279
50	1-propanol	C_3H_8O	60.1	562.7	124.462	1.25704
51	2-methylbutan-1-ol	$C_5H_{12}O$	88.1	736	206.685	1.24314
52	Isopropyl acetate	$C_5H_{10}O_2$	102.1	657.1	164.2	1.16453
53	pentanal	$C_5H_{10}O$	86.1	690.5	178.596	1.18696

As shown in Figure 8, each row of the figure indicates all the signals selected from the cucumber samples, and each column indicate the signal peak intensity of the same volatile component in cucumber samples with different storage times for different treatments. As can be seen in Figure 8, the volatile flavor substances in cucumbers of different treatments changed with the extension of storage time for each treatment. The contents of trans-2-hexenal, hydroxyacetone, ethyl acetate, ethylene glycol dimethyl ether, glutaraldehyde, isopropyl acetate, hexanal, 3-methylbutyraldehyde, benzaldehyde, and trans-cis-2,6-nonadienal showed different degrees of decrease. The aldehydes in the cucumber showed different degrees of decline with storage time, indicating a decrease in aroma significantly, probably due to microbial growth. These results indicated that the cucumber in each treatment showed different degrees of aroma loss during storage, while the aldehyde content of the cucumber in CS+P treatment groups decreased slowly compared to the NCS+P treatment, indicating that the cold stimulation combined with perforated spontaneous air conditioning packaging treatment helped the cucumber flavor retention during storage, where CS20+P and CS40+P treatments were better than the other two treatment groups. The concentrations of 2-ethyl-5-methylpyrazine, nonanal, cis-3-hexenol, (E)-2-heptenal, trans-2,4-heptadienal, octanal, trans-2-hexenal, heptanal, 2-propanol, isobutanol, 3-hydroxy-2-butanone, and n-hexanol first increased and then decreased. The reason for this change in the aldehydes of C6–C9 (the main aroma of cucumber) may be that cucumber had not yet reached full maturity at the beginning of storage, and its main aroma had not yet reached its peak, and then declined after reaching the peak perhaps due to microorganisms and the aging of the cucumber itself. The concentration of 3-pentanone, methallyl sulfide, phenylethanal, 2,3-butanediol, 1-propanol, 2-isobutyric acid, caprylic acid, 2-hexanol, methyl butyrate, 2-methyl-1-butanol, butyraldehyde, propionic acid, N,N-diethyl ethylamine, ethyl butyrate, and 3-butenenitrile in NCS+P group increased with storage time, which also may be as a result of microbial and self-metabolic influences. The high content of alcohols may be due to the reaction between acyl-CoA and alcohols, microbial action, and oxidative aging of fruits, catalyzed by ester alcohol-acyltransferase [49], as well as the high content of amines and acids with unpleasant odors, etc., indicating that the effect of CS+P treatment on flavor retention of cucumber was better than that of NCS+P treatment. In conclusion, CS combined with PM-PMAP treatment had a good effect on cucumber flavor retention, and CS20+P and CS40+P treatment groups had the best effect.

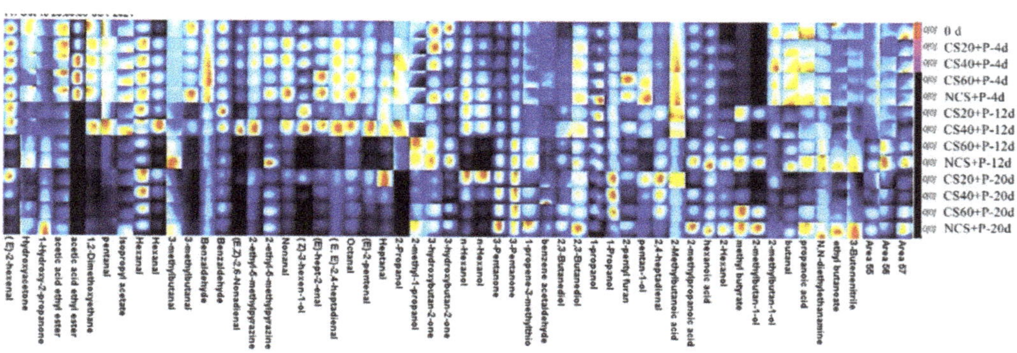

Figure 8. Fingerprints of volatile organic compounds in cucumbers under different treatments. The color depth represents the concentrations of the volatile compounds. Red indicates that the concentration of the substance is high, while blue indicates that the concentration of the substance is low. CS20+P—packaged after 20 min of cold shock; CS40+P—packaged after 40 min of cold shock; CS60+P—packaged after 60 min of cold shock; NCS+P—packaged after 0 min of cold shock.

PCA analysis is a method to reveal the intrinsic relationship between multivariate data or samples using the idea of dimensionality reduction to use a small number of integrated variables to replace the original complex multivariate quantities, to reduce data complexity [50], and to visualize the data. As shown in Figure 9, the contribution of the first principal component was 42.68%, and the contribution of the second principal component was 21.53%; the cumulative contribution of these two principal components was 64.21%, which represented most of the information of the original data. The relative proximity of the samples within the group indicated good reproducibility of the samples. The cucumbers of the four treatments were distributed in different areas, indicating that there were differences in VOCs among the samples, and the flavor of the cucumbers of the four treatments was changed more obviously with the extension of storage time. The flavor of cucumbers in all treatment groups on day 20 was more different compared with day 0. The flavor of cucumbers in the NCS+P treatment group changed most on day 20. The flavor of cucumbers in the CS20+P and CS40+P treatment groups at the end of storage was less different compared to day 0, indicating that CS20+P and CS40+P had a good effect on cucumber flavor retention.

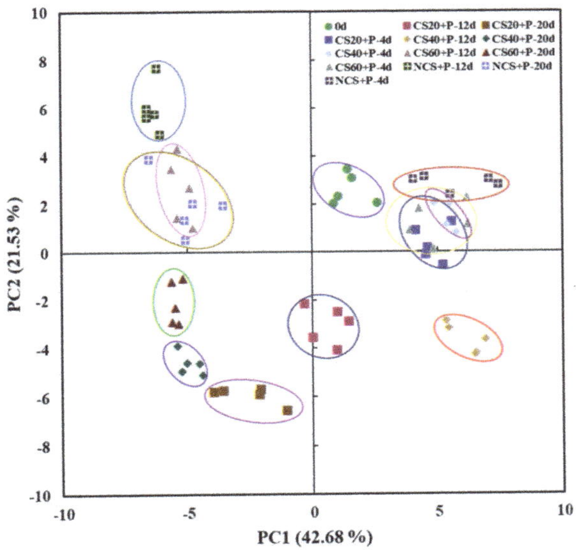

Figure 9. PCA scores of volatile organic compounds in cucumbers under different treatments. CS20+P—packaged after 20 min of cold shock; CS40+P—packaged after 40 min of cold shock; CS60+P—packaged after 60 min of cold shock; NCS+P—packaged after 0 min of cold shock.

PLS–DA analysis is similar to PCA analysis in that it also performs a dimensionality reduction analysis method, but it can be pre-classified, which can remove the possible influence of uncontrolled variables on the data analysis [51], further mining the information in the data and quantifying the extent to which characteristic compounds cause component differences. As shown in Figure 10, after PLS–DA analysis, R^2X was obtained as 0.976, R^2Y as 0.851, and Q^2 as 0.580, with the X matrix as the variable (VOCs) observation matrix, and the Y matrix as the category (cucumbers after different treatment) attribution matrix; here, R^2X and R^2Y indicate the percentages of X and Y matrix information explained by the model, respectively; Q^2 represents the prediction rate of the model [52]. These indicated a model with a good stability and a good predictability for cucumbers with different treatments. It can be used for discriminant analysis of cucumber samples during storage.

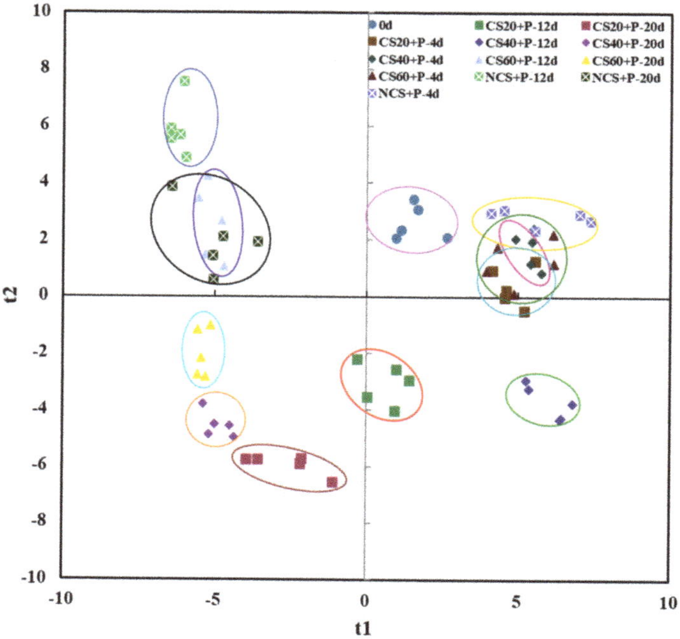

Figure 10. PLS–DA scores of volatile organic compounds in cucumbers under different treatments. CS20+P—packaged after 20 min of cold shock; CS40+P—packaged after 40 min of cold shock; CS60+P—packaged after 60 min of cold shock; NCS+P—packaged after 0 min of cold shock.

Variable importance for the projection (VIP) can quantify the contribution of each variable of PLS–DA to classification. It is usually considered that VIP greater than 1 indicates an important role in the discriminatory process; and the larger the VIP value, the more significant the difference of variables among cucumbers of different treatments [53]. A panel of 20 VOCs with p-value < 0.05 and VIP score > 1 were selected as potential markers for the discrimination of cucumbers with different treatments. These 20 VOCs were 2-ethyl-5-methyl pyrazine, isobutyric acid, 2,3-butanediol, isobutanol, 1-propanol, benzaldehyde phenylacetaldehyde, 3-butenenitrile, 3-methylbutyraldehyde, 2-pentylfuran, 2-methyl-1-butanol, 2-hexanol, trans-2,4-heptadienal, trans, cis-2,6-nonadienal, 1-pentanol, hexanoic acid, ethyl acetate, ethyl butyrate, 3-pentanone, and methyl butyrate. The largest contributing VOC was 2-ethyl-5-methyl pyrazine, which was particularly important in distinguishing between different treatments of cucumber.

4. Conclusions

The data presented in this article show that the combination of CS and PM-PMAP treatment was beneficial to maintaining the quality of cucumber after harvest, such as minimal weight loss, color change, hardness, and total soluble solids reduction, as well as ascorbic acid retention, and malondialdehyde increment. Similarly, flavor change and respiratory rate were minimized. In summary, when the cold shock time was 40 min, the preservation quality of cucumber was the best. As a simple, safe, and inexpensive post-harvest processing technology, the use of CS combined with PM-PMAP to manipulate physiological and biochemical activities was found to be a practical method to extend the shelf life of cucumber fruits during the storage period. Future research may include different types of fruits and vegetables using the combined pretreatment method. The optimal processing parameters (cold shock intensity, time, and packaging materials) can be explored through comprehensive experiments.

Author Contributions: Conceptualization, F.W. and S.M.; methodology, F.W. and S.M.; validation, F.W. and S.M.; formal analysis, F.W.; investigation, F.W.; resources, J.L.; data curation, F.W.; writing—original draft preparation, F.W.; writing—review and editing, B.C. and X.W.; visualization, F.W.; project administration, X.W. All authors have read and agreed to the published version of the manuscript.

Funding: This research was funded by the Industrial Innovation Team Program of Vegetables of Modern Agricultural Industry Technology System of Hebei Province [Grant Number HBCT2018030208].

Institutional Review Board Statement: Not applicable.

Informed Consent Statement: Not applicable.

Data Availability Statement: The data presented in this study are available in article.

Acknowledgments: The authors would like to acknowledge Hebei Agricultural University for providing instruments used in the experiments.

Conflicts of Interest: The authors declare no conflict of interest.

References

1. Mahajan, P.V.; Caleb, O.J.; Gil, M.I.; Izumi, H.; Colelli, G.; Watkins, C.B.; Zude, M. Quality and safety of fresh horticultural commodities: Recent advances and future perspectives. *Food Packag. Shelf Life* **2017**, *14*, 2–11. [CrossRef]
2. Li, M.; Yu, H.; Xie, Y.; Guo, Y.; Cheng, Y.; Qian, H.; Yao, W. Effects of double layer membrane loading eugenol on postharvest quality of cucumber. *LWT-Food Sci. Technol.* **2021**, *145*, 111310. [CrossRef]
3. Jia, B.; Zheng, Q.; Zuo, J.; Gao, L.; Wang, Q.; Guan, W.; Shi, J. Application of postharvest putrescine treatment to maintain the quality and increase the activity of antioxidative enzyme of cucumber. *Sci. Hortic.* **2018**, *239*, 210–215. [CrossRef]
4. Tarek, A.R.; Rasco, B.A.; Sablani, S.S. Ultraviolet-C light sanitization of english cucumber (*Cucumis sativus*) packaged in polyethylene film. *J. Food Sci.* **2016**, *81*, E1419–E1430. [CrossRef] [PubMed]
5. Nasef, I.N. Short hot water as safe treatment induces chilling tolerance and antioxidant enzymes, prevents decay and maintains quality of cold-stored cucumbers. *Postharvest Biol. Technol.* **2018**, *138*, 1–10. [CrossRef]
6. Patel, C.; Panigrahi, J. Starch glucose coating-induced postharvest shelf-life extension of cucumber. *Food Chem.* **2019**, *288*, 208–214. [CrossRef]
7. Li, J.; Li, Q.; Lei, X.; Tian, W.; Cao, J.; Jiang, W.; Wang, M. Effects of wax coating on the moisture loss of cucumbers at different storage temperatures. *J. Food Qual.* **2018**, *2018*, 9351821. [CrossRef]
8. Manjunatha, M.; Anurag, R.K. Effect of modified atmosphere packaging and storage conditions on quality characteristics of cucumber. *J. Food Sci. Technol.* **2014**, *51*, 3470–3475. [CrossRef] [PubMed]
9. Feng, L.; Zhang, M.; Adhikari, B.; Guo, Z. Effect of ultrasound combined with controlled atmosphere on postharvest storage quality of cucumbers (*Cucumis sativus* L.). *Food Bioprocess Technol.* **2018**, *11*, 1328–1338. [CrossRef]
10. Lwin, W.W.; Pongprasert, N.; Boonyaritthongchai, P.; Wongs-Aree, C.; Srilaong, V. Synergistic effect of vacuum packaging and cold shock reduce lignification of asparagus. *J. Food Biochem.* **2020**, *44*, e13479. [CrossRef]
11. Gu, S.; Xu, D.; Zhou, F.; Gao, H.; Hu, W.; Gao, X.; Jiang, A. Cold shock treatment maintains quality and induces relative expression of cold shock domain protein (*CSDPs*) in postharvest sweet cherry. *Sci. Hortic.* **2020**, *262*, 109058. [CrossRef]
12. Zhang, H.; Yang, S.; Joyce, D.C.; Jiang, Y.; Qu, H.; Duan, X. Physiology and quality response of harvested banana fruit to cold shock. *Postharvest Biol. Technol.* **2010**, *55*, 154–159. [CrossRef]
13. Chen, J.; Liu, X.; Li, F.; Li, Y.; Yuan, D. Cold shock treatment extends shelf life of naturally ripened or ethylene-ripened avocado fruits. *PLoS ONE* **2017**, *12*, e0189991. [CrossRef]
14. Karagöz, Ş.; Demirdöven, A. Effect of chitosan coatings with and without Stevia rebaudiana and modified atmosphere packaging on quality of cold stored fresh-cut apples. *LWT-Food Sci. Technol.* **2019**, *108*, 332–337. [CrossRef]
15. Sandhya. Modified atmosphere packaging of fresh produce: Current status and future needs. *LWT-Food Sci. Technol.* **2010**, *43*, 381–392. [CrossRef]
16. Olawuyi, I.F.; Park, J.J.; Lee, J.J.; Lee, W.Y. Combined effect of chitosan coating and modified atmosphere packaging on fresh-cut cucumber. *Food Sci. Nutr.* **2019**, *7*, 1043–1052. [CrossRef] [PubMed]
17. Paulsen, E.; Barrios, S.; Lema, P. Ready-to-eat cherry tomatoes: Passive modified atmosphere packaging conditions for shelf life extension. *Food Packag. Shelf Life* **2019**, *22*, 100407. [CrossRef]
18. Fan, K.; Zhang, M.; Guo, C.; Dan, W.; Devahastin, S. Laser-induced microporous modified atmosphere packaging and chitosan carbon-dot coating as a novel combined preservation method for fresh-cut cucumber. *Food Bioprocess Technol.* **2021**, *14*, 968–983. [CrossRef]
19. Jin, C.; Cai, Z. A circular arc approximation algorithm for cucumber classification with image analysis. *Postharvest Biol. Technol.* **2020**, *165*, 111184. [CrossRef]
20. Dhall, R.K.; Sharma, S.R.; Mahajan, B.V. Effect of shrink wrap packaging for maintaining quality of cucumber during storage. *J. Food Sci. Technol.* **2012**, *49*, 495–499. [CrossRef]

21. Yang, L.; Zhang, C.; Cong, P.; Cheng, Y.; Wang, Q. Texture parameters of different apple varieties' flesh as measured by texture profile analysis. *Food Sci.* **2014**, *35*, 57–62. [CrossRef]
22. Yang, Z.; Cao, S.; Cai, Y.; Zheng, Y. Combination of salicylic acid and ultrasound to control postharvest blue mold caused by Penicillium expansum in peach fruit. *Innov. Food Sci. Emerg. Technol.* **2011**, *12*, 310–314. [CrossRef]
23. Wang, B.; Zhu, S. Pre-storage cold acclimation maintained quality of cold-stored cucumber through differentially and orderly activating ROS scavengers. *Postharvest Biol. Technol.* **2017**, *129*, 1–8. [CrossRef]
24. Li, M.; Yang, R.; Zhang, H.; Wang, S.; Chen, D.; Lin, S. Development of a flavor fingerprint by HS-GC-IMS with PCA for volatile compounds of Tricholoma matsutake Singer. *Food Chem.* **2019**, *290*, 32–39. [CrossRef] [PubMed]
25. Song, J.; Shao, Y.; Yan, Y.; Li, X.; Peng, J.; Guo, L. Characterization of volatile profiles of three colored quinoas based on GC-IMS and PCA. *LWT-Food Sci. Technol.* **2021**, *146*, 111292. [CrossRef]
26. Choi, D.S.; Park, S.H.; Choi, S.R.; Kim, J.S.; Chun, H.H. The combined effects of ultraviolet-C irradiation and modified atmosphere packaging for inactivating Salmonella enterica serovar Typhimurium and extending the shelf life of cherry tomatoes during cold storage. *Food Packag. Shelf Life* **2015**, *3*, 19–30. [CrossRef]
27. Nasiri, M.; Barzegar, M.; Sahari, M.A.; Niakousari, M. Tragacanth gum containing *Zataria multiflora* Boiss. essential oil as a natural preservative for storage of button mushrooms (*Agaricus bisporus*). *Food Hydrocoll.* **2017**, *72*, 202–209. [CrossRef]
28. Wei, W.; Lv, P.; Xia, Q.; Tan, F.; Sun, F.; Yu, W.; Jia, L.; Cheng, J. Fresh-keeping effects of three types of modified atmosphere packaging of pine-mushrooms. *Postharvest Biol. Technol.* **2017**, *132*, 62–70. [CrossRef]
29. Azevedo, S.; Cunha, L.M.; Oliveira, J.C.; Mahajan, P.V.; Fonseca, S.C. Modelling the influence of time, temperature and relative humidity conditions on the mass loss rate of fresh oyster mushrooms. *J. Food Eng.* **2017**, *212*, 108–112. [CrossRef]
30. Zhao, X.; Xia, M.; Wei, X.; Xu, C.; Luo, Z.; Mao, L. Consolidated cold and modified atmosphere package system for fresh strawberry supply chains. *LWT-Food Sci. Technol.* **2019**, *109*, 207–215. [CrossRef]
31. Zhao, S.; Yang, Z.; Zhang, L.; Luo, N.; Wang, C. Effects of different direction of temperature jump treatment on cucumbers. *J. Food Process Eng.* **2018**, *41*, e12600. [CrossRef]
32. Chen, C.; Zhou, G.; Chen, J.; Liu, X.; Lu, X.; Chen, H.; Tian, Y. Integrated metabolome and transcriptome analysis unveils novel pathway involved in the formation of yellow peel in cucumber. *Int. J. Mol. Sci.* **2021**, *22*, 1494. [CrossRef] [PubMed]
33. Cardenas-Coronel, W.G.; Carrillo-Lopez, A.; Velez de la Rocha, R.; Labavitch, J.M.; Baez-Sanudo, M.A.; Heredia, J.B.; Zazueta-Morales, J.J.; Vega-Garcia, M.O.; Sanudo-Barajas, J.A. Biochemistry and cell wall changes associated with noni (*Morinda citrifolia* L.) fruit ripening. *J. Agric. Food Chem.* **2016**, *64*, 302–309. [CrossRef]
34. Maleki, G.; Sedaghat, N.; Woltering, E.J.; Farhoodi, M.; Mohebbi, M. Chitosan-limonene coating in combination with modified atmosphere packaging preserve postharvest quality of cucumber during storage. *J. Food Meas. Charact.* **2018**, *12*, 1610–1621. [CrossRef]
35. Fan, K.; Zhang, M.; Fan, D.; Jiang, F. Effect of carbon dots with chitosan coating on microorganisms and storage quality of modified-atmosphere-packaged fresh-cut cucumber. *J. Sci. Food Agric.* **2019**, *99*, 6032–6041. [CrossRef] [PubMed]
36. Wei, Y.; Liu, Z.; Su, Y.; Liu, D.; Ye, X. Effect of salicylic acid treatment on postharvest quality, antioxidant activities, and free polyamines of asparagus. *J. Food Sci.* **2011**, *76*, S126–S132. [CrossRef]
37. Maurya, V.K.; Ranjan, V.; Gothandam, K.M.; Pareek, S. Exogenous gibberellic acid treatment extends green chili shelf life and maintain quality under modified atmosphere packaging. *Sci. Hortic.* **2020**, *269*, 108934. [CrossRef]
38. Valverde-Miranda, D.; Díaz-Pérez, M.; Gómez-Galán, M.; Callejón-Ferre, Á.-J. Total soluble solids and dry matter of cucumber as indicators of shelf life. *Postharvest Biol. Technol.* **2021**, *180*, 111603. [CrossRef]
39. Barman, K.; Siddiqui, M.W.; Patel, V.B.; Prasad, M. Nitric oxide reduces pericarp browning and preserves bioactive antioxidants in litchi. *Sci. Hortic.* **2014**, *171*, 71–77. [CrossRef]
40. Hussein, Z.; Caleb, O.J.; Opara, U.L. Perforation-mediated modified atmosphere packaging of fresh and minimally processed produce—A review. *Food Packag. Shelf Life* **2015**, *6*, 7–20. [CrossRef]
41. Fan, K.; Zhang, M.; Bhandari, B.; Jiang, F. A combination treatment of ultrasound and ε-polylysine to improve microorganisms and storage quality of fresh-cut lettuce. *LWT-Food Sci. Technol.* **2019**, *113*, 108315. [CrossRef]
42. Chitravathi, K.; Chauhan, O.P.; Raju, P.S. Influence of modified atmosphere packaging on shelf-life of green chillies (*Capsicum annuum* L.). *Food Packag. Shelf Life* **2015**, *4*, 1–9. [CrossRef]
43. Reche, J.; García-Pastor, M.E.; Valero, D.; Hernández, F.; Almansa, M.S.; Legua, P.; Amorós, A. Effect of modified atmosphere packaging on the physiological and functional characteristics of Spanish jujube (*Ziziphus jujuba* Mill.) cv 'Phoenix' during cold storage. *Sci. Hortic.* **2019**, *258*, 108743. [CrossRef]
44. Hashemi, S.M.B. Effect of pulsed ultrasound treatment compared to continuous mode on microbiological and quality of Mirabelle plum during postharvest storage. *Int. J. Food Sci. Technol.* **2018**, *53*, 564–570. [CrossRef]
45. Sothornvit, R.; Kiatchanapaibul, P. Quality and shelf-life of washed fresh-cut asparagus in modified atmosphere packaging. *LWT-Food Sci. Technol.* **2009**, *42*, 1484–1490. [CrossRef]
46. Zheng, Q.; Zuo, J.; Gu, S.; Gao, L.; Hu, W.; Wang, Q.; Jiang, A. Putrescine treatment reduces yellowing during senescence of broccoli (*Brassica oleracea* L. var. *italica*). *Postharvest Biol. Technol.* **2019**, *152*, 29–35. [CrossRef]
47. Zhao, S.; Yang, Z.; Zhang, L.; Luo, N.; Li, X. Effect of combined static magnetic field and cold water shock treatment on the physicochemical properties of cucumbers. *J. Food Eng.* **2018**, *217*, 24–33. [CrossRef]

48. Chen, Y.P.; Cai, D.; Li, W.; Blank, I.; Liu, Y. Application of gas chromatography-ion mobility spectrometry (GC-IMS) and ultrafast gas chromatography electronic-nose (uf-GC E-nose) to distinguish four Chinese freshwater fishes at both raw and cooked status. *J. Food Biochem.* **2021**, *00*, e13840. [CrossRef]
49. YAN, T.; QIN, H.; ZHANG, P.; TIAN, S.; LI, J.; LI, B. Effects of 1-Methylcyclopropene combined with ξ-Polylysine on quality and volatile components of fuji apples during shelf life after cold storage. *Food Sci.* **2018**, *39*, 207–214. [CrossRef]
50. He, J.; Wu, X.; Yu, Z. Microwave pretreatment of camellia (*Camellia oleifera* Abel.) seeds: Effect on oil flavor. *Food Chem.* **2021**, *364*, 130388. [CrossRef]
51. Ruiz-Perez, D.; Guan, H.; Madhivanan, P.; Mathee, K.; Narasimhan, G. So you think you can PLS-DA? *BMC Bioinform.* **2020**, *21*, 2. [CrossRef] [PubMed]
52. Li, D.; Peng, J.; Kwok, L.; Zhang, W.; Sun, T. Metabolomic analysis of *Streptococcus thermophilus* S10-fermented milk. *LWT-Food Sci. Technol.* **2022**, *161*, 113368. [CrossRef]
53. Farres, M.; Pina, B.; Tauler, R. Chemometric evaluation of Saccharomyces cerevisiae metabolic profiles using LC-MS. *Metabolomics* **2015**, *11*, 210–224. [CrossRef] [PubMed]

foods

MDPI

Article

Chlorophyllin-Based 405 nm Light Photodynamic Improved Fresh-Cut Pakchoi Quality at Postharvest and Inhibited the Formation of Biofilm

Yuchen Zhang [1,2], Zhaoyang Ding [1,2], Changbo Shao [3] and Jing Xie [1,2,4,*]

1 College of Food Science and Technology, Shanghai Ocean University, Shanghai 201306, China
2 National Experimental Teaching Demonstration Center for Food Science and Engineering, Shanghai Ocean University, Shanghai 201306, China
3 School of Intelligent Manufacturing and Service, Shandong Institute of Commerce and Technology, Jinan 250103, China
4 Shanghai Professional Technology Service Platform on Cold Chain Equipment Performance and Energy Saving Evaluation, Shanghai 201306, China
* Correspondence: jxie@shou.edu.cn; Tel.: +86-0216-190-0391

Abstract: The aim of this study was to evaluate the effect of chlorophyllin-based photodynamic inactivation (Chl-PDI) on biofilm formation and fresh-cut pakchoi quality during storage. Firstly, Chl-based PDI reduced the amount of biofilm in an in vivo experiment and inactivated the food spoilage bacteria. Antibacterial mechanism analysis indicated that the bacterial extracellular polysaccharides and extracellular proteins were vulnerable targets for attacks by the Chl-based PDI. Then, the food spoilage microorganisms (*Pseudomonas reinekei* and *Pseudomonas palleroniana*) were inoculated onto the surface of fresh-cut pakchoi. We used chlorophyllin (1×10^{-5} mol/L) and 405 nm light (22.27 J/cm^2 per day) to investigate the effect of Chl-based PDI treatment on fresh-cut pakchoi quality during storage. The results showed that Chl-based PDI increased the visual quality and the content of chlorophyll, VC, total soluble solids, and SOD activity and decreased the occurrence of leaf yellowing and POD activity. These suggest that Chl-based PDI can be used for the preservation of fresh-cut pakchoi and has the potential to inhibit biofilm formation of food spoilage bacteria. It is of great significance for the effective processing and traditional vegetable preservation.

Keywords: photodynamic inactivation (PDI); antimicrobial; biofilm; fresh-cut pakchoi; preservation

Citation: Zhang, Y.; Ding, Z.; Shao, C.; Xie, J. Chlorophyllin-Based 405 nm Light Photodynamic Improved Fresh-Cut Pakchoi Quality at Postharvest and Inhibited the Formation of Biofilm. *Foods* **2022**, *11*, 2541. https://doi.org/10.3390/foods11162541

Academic Editor: Peng Jin

Received: 19 July 2022
Accepted: 15 August 2022
Published: 22 August 2022

1. Introduction

Nowadays, fresh-cut vegetables are expanding sharply in the global market [1]. However, fresh-cut vegetables are susceptible to contamination by microorganisms during transportation, storage, distribution, and consumption, which results in spoilage and deterioration [1,2]. The Food and Agriculture Organization of the United Nations (FAO) reported that approximately 14% of food loss happens before the retail stage [3]. However, about 9–20% of fruits and vegetables are wasted in the consumption stage [3]. Studies have shown that *Pseudomonas* spp. cause various kinds of fresh-cut vegetables to spoil [4–6]. This is evident in cases in which *Pseudomonas* spp. caused lettuce to rot [5] while *Pseudomonas cichorii* resulted in the spoilage of and necrotic spots on heads of lettuce [6]. Another example of *Pseudomonas* spp.-caused spoilage was when pectolytic enzymes produced by *Pseudomonas viridiflava* and *Pseudomonas chlororaphis* accelerated the rate of rot and caused soft rot of vegetables [7,8]. Our previous work found that *Pseudomonas reinekei* and *Pseudomonas palleroniana* showed highly the dominant decaying bacteria of fresh-cut pakchoi [9,10]. Moreover, Morris et al. [11] found that bacteria formed biofilms easily on the surface of vegetables, such as parsley and lettuce, and that the bacterial biofilms formed on the surface of vegetables are difficult to remove, which adversely affects the freshness storage of vegetables. Biofilms enhance bacterial resistance and promote the spread of bacteria, causing great harm to the safety of vegetables [12]. To reduce the microbial contamination, a non-thermal sterilization process, such as UV, sodium hypochlorite, 405 nm light, etc., was used. However, the excessive use of chlorine can be carcinogenic [13] and is forbidden in food industries in Sweden, Germany, Belgium, and other European countries [14]. Additionally, excessive UV exposure is harmful to human skin and eyes and 405 nm light sterilization efficiency is low [10]. Hence, there is a need to develop a new sterilization technology for the fresh-cut vegetable industry.

Plant-based photodynamic inactivation (PDI), which showed high sterilization efficiency, might serve as a promising answer [15–17]. PDI is a new, non-thermal processing method used in food and medicine. Chlorophyllin (Chl) is a semi-synthetic porphyrin, water-soluble food colorant, also known as food additive E140 [18]. After being activated by 405 nm light, it produces reactive oxygen species (ROS) around the surface, causing lethal damage to bacteria, viruses, and fungi [19,20]. Studies have shown the potential of PDI in vegetable preservation without a negative impact on quality. This is exemplified in the work that showed that a chlorophyllin photosensitization treatment prolonged the shelf life of strawberries for 3 days and maintained the antioxidant activity and color of strawberries [17]. In a similar case, Paskeviciute et al., (2018) [21] found that Chl-based photosensitization reduced 2.4 log CFU/mL microbiota on tomatoes. Hence, plant-based PDI may open a new way for the development of non-thermal, effective, and eco-friendly antibacterial technology [22]. Additionally, some studies reported that PDI effectively inhibits the formation of biofilm [23,24]. For example, Chen et al., (2020) [24] used curcumin-based PDI eradiated 90% biofilm and reduced the key components of the extracellular polymers of *Vibrio parahaemolyticus*. Another example, in the work undertaken by Silva et al., (2018) [25], showed that rose bengal and erythrosine-based PDI inhibited the biofilm formation of *Staphylococcus aureus* and *Listeria innocua*. Bonifácio et al., (2018) [26] found that curcumin-based, blue-light PDI reduced 4.9 log of viable biofilm of *Listeria innocua*. However, inhibition by Chl-based PDI of biofilm formation on vegetables has been rarely reported.

The aim of this study was to evaluate the antimicrobial efficiency of Chl-based PDI against *Pseudomonas reinekei* and *Pseudomonas palleroniana*. We investigated the effect of Chl-based PDI on the quality of fresh-cut pakchoi during the storage stage by measuring organoleptic properties, color parameters, weight loss, water distribution and migration, soluble solids, chlorophyll content, VC content, superoxide dismutase (SOD) content, and peroxidase (POD) content. Moreover, the inhibition of biofilm by Chl-based PDI was also investigated by confocal laser scanning microscope (CLSM), extracellular polysaccharide, and lipid assays.

2. Materials and Methods

2.1. LED Lightbox Setup

An LED illumination system was set in a refrigerator (BCD-252MHV, SSEC, Suzhou, China) with high-intensity 405 nm LED lamps (15.12 μmol/(m^2·s), 5.1 W/m^2, WAN-T8120, WEGA, Qingzhou, China). The wavelength range of visible light was between 400 and 780 nm. The irradiance of 405 nm LED through the PVC packaging materials at 30 cm (vertical distance) was 5.1 w/m^2, which was measured by an LED radiometer (ST-513, SENTRY, Taipei, Taiwan).

2.2. Culture of Bacterial Strains and the Preparation of Cocktail Solutions

In our previous work, we selected two types of specific spoilage microorganisms (SSOs), *Pseudomonas reinekei* MT1 (*P. reinekei*) and *Pseudomonas palleroniana* (*P. palleroniana*) CFBP4389 [9,10]. *P. reinekei* and *P. palleroniana* were isolated from spoiled pakchoi and caused severe spoilage when they were inoculated with fresh pakchoi. Details of the culture of bacterial strains were described by Zhang et al. [27]. *P. reinekei* and *P. palleroniana* were prepared with a final concentration of approximately 10^6 CFU/mL.

2.3. Chemicals and Treatment Method of Pakchoi

Chlorophyll sodium salt (Chl) was obtained from Yuanye Bio-Technology Co., Ltd., Shanghai, China. The solution was prepared with 1.5×10^{-5} M of chlorophyllin. Pakchoi plants were grown on a commercial farm in Pudong, Shanghai, China. Samples were sterilized by 200 ppm of NaClO solution and dried. After that, samples were cut by a sterile scalpel 1 cm away from the root. Then, as Figure 1 shows, the leaves were soaked in the cell suspensions (10^6 CFU/mL) for 5 min and dried again. The control group was treated without light or Chl. The Chl group was treated with Chl. The light group was treated with light. The Chl+light group was treated with light and Chl. A solution of Chl was sprayed on the surface of leaves of the Chl and Chl+light groups and dried before storage (spray: 5 g of solution per 100 g of sample). Additionally, the same amount of sterile water was sprayed on the leaves of the pakchoi of the light and control groups. Then, the 405 nm LED illumination system provided 12 h of irradiation (22.27 J/cm^2) every day at 4 °C. Every sample was collected at 0 d, 2 d, 4 d, 6 d, 8 d, 10 d, and 12 d (a total dose of 0, 44.54, 89.08, 133.63, 178.17, 222.71, and 267.25 J/cm^2, respectively).

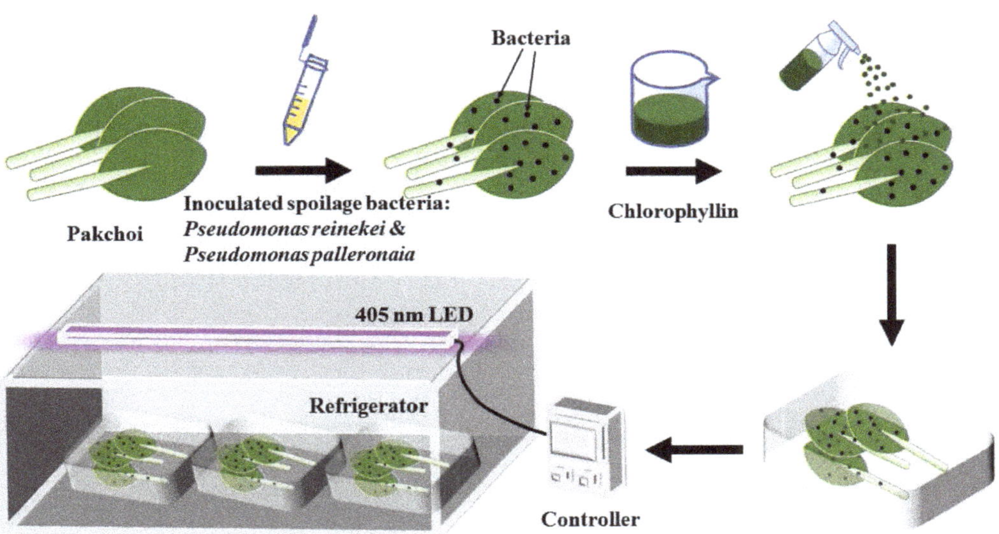

Figure 1. Schematic diagram of the Chl+light group experimental arrangement.

2.4. Organoleptic Properties, Color Parameters, Chlorophyll Content, and Weight Loss Rate

The methods of organoleptic properties were according to Zhang et al. [9]. The panel was made up of 20 dietetic students and three food science faculty members, who conducted sensory evaluations of pakchoi samples. These students were enrolled in graduate research and experimental foods courses, where they gained experience in translating their perceptions of color, smell, and form into descriptive words and numbers on scoring sheets. The panel that conducted a sensory evaluation of samples used a 10-grade marking system: 10.0–9.0 (excellent), 8.9–7.0 (good), 6.9–5.0 (fair), and 4.9–0 (discardable). The surface color of pakchoi was determined using a CR-400 Chroma Meter (Konica Minolta Sensing Inc., Osaka, Japan). In order to have homogeneous color samples, the measurements used were only obtained from the green part of the pakchoi leaf. The color coordinates ranged from L* = 0 (black) to L* = 100 (white), −a* (greenness) to +a* (redness), and –b* (blueness) to +b* (yellowness).

The methods of determining the chlorophyll content were according to Hasperué et al. [28] with some modifications. A total of 5 g of leaf tissue of pakchoi was homogenized in 20 mL of 80% acetone with a tissue homogenizer at 2000× g for 30 s. The absorbance of the filtered homogenate was measured in UV-1102 (Tianmei Instrument Co., Ltd., Changsha, China) at 645 and 663 nm. The weight loss (%) was calculated using the following equation:

$$X = \frac{W_0 - W_1}{W_0} \times 100\% \quad (1)$$

where X = weight loss rate (%), W_0 = the weight of pakchoi on day 0 in g, and W_1 = the weight of pakchoi with different treatments in g.

2.5. Soluble Solids, Water Distribution and Migration, VC Content, Superoxide Dismutase (SOD), and Oxidase (POD) Content

The methods of determining the soluble solids and water distribution and migration content were according to Zhang et al., (2021) [9]. We took 5 g of a sample and fully grinded and centrifugated it for 10 min at 3500 rpm. Then, we took a drop of the supernatant on the inspection mirror of a digital refractometer (PR32a, ATAGO, Japan) and recorded the reading. The pakchoi samples were cut into 1 cm × 1 cm pieces and tested by LF-NMR (0.5 T, 23.2 MHz, PQ001, Niumag Electric Co., Shanghai, China) for water distribution and migration testing. Acquisition parameters were as follows: coil temperature = 32 °C, proton resonance frequency = 24 MHz, CPMG sequence was used, SW (sampling frequency) = 100 Hz, RG_1 (analog gain) = 20, P_1 = 20.00 μs, DRG1 (digital gain) = 3, TD = 1024, PRG1 = 3, TW (repeated sampling times) = 15,000, NS (accumulation times) = 4, P = 35 μs, TE (echo time) = 0.500, and echo number (nech) = 3000. The T_2 spectrum was obtained by iterative inversion using the analysis software provided by Niumag Electric Co., Ltd., Shanghai, China. The methods of determining the VC content were according to Wang et al. [29]. The leaf tissue (1.0 g) of samples was ground in 5 mL 0.05 mol L^{-1} oxalic acid-0.2 mM EDTA. The supernatant was collected by centrifuging at 13,000× *g* for 20 min. The methods of determining the SOD and POD activity were according to He et al. [30]. The reaction mixture consisted of a supernatant (0.5 mL), 0.1 mol L^{-1} PBS (1 mL, pH 7.8), 0.2 % (*v*/*v*, 0.9 mL) guaiacol, and 0.3% (*v*/*v*, 0.6 mL) H_2O_2.

2.6. *Pseudomonas* spp. Count, Extracellular Polysaccharide, Extracellular Protein, and Attached Biomass

The *Pseudomonas* spp. count was determined according to Federico et al. [5]. We weighed 10 g of fresh-cut pakchoi samples in the aseptic operation platform, then put 90 mL 0.1 M PBS into an aseptic homogenizing bag, and made a 1:10 sample homogenization solution. The CFC (Qingdao Haibo Biotechnology Co., Ltd., Qingdao, China) was used to culture and determine *Pseudomonas* spp. count.

Sterile stainless steel sheets were added to the LB medium supplemented with bacteria; then, light (22.27 J/cm^2 per day) or Chl (1.5 × 10^{-5} M) treatment was used. The no light and Chl treatment group was used as the control. Biofilm extracellular polysaccharides and proteins were measured after a standing culture for 0 d, 6 d, and 12 d at 4 °C. The biofilm was peeled off into sterile water by ultrasound; then, pH was adjusted to neutral. The extracellular polysaccharides' content and extracellular protein were measured by the phenol-sulfuric acid method and coomassie brilliant blue method, respectively [31]. The biofilm biomass was quantified using crystal violet staining [32].

2.7. Confocal Laser Scanning Microscope (CLSM)

Samples were washed with deionized water to remove unattached bacteria. Then, the samples were stained with SYBR Green I fluorescence. A CLMS machine (LSM710, Carl Zeiss, Oberkochen, Germany) was used to monitor samples that excited at 488 nm and emitted at 500–550 nm [24].

2.8. Data Analysis Method

All experiments were performed for at least three independent trials. All data were expressed as mean ± standard error (*n* = 3) and performed by one-way analysis of variance

(ANOVA). Additionally, the differences among the means were compared by Duncan's multiple range test with a significance of $p < 0.05$ using the SPSS 22.0 statistical program.

3. Results and Discussion

3.1. Inhibition of Chl-Based PDI against Inoculated Spoilage Microorganisms

In the present study, the inhibitory effect of Chl-based PDI on inoculated spoilage microorganisms was investigated with different treatments (Figure 2d). The initial populations of colonies on samples were 3.68 ± 0.15 log CFU/g after inoculation. However, the bacteria in the control and Chl groups increased rapidly from day 4. By day 10, the populations of microorganisms in the control and Chl groups reached to 6.83 ± 0.18 log CFU/g and 5.29 ± 0.28 log CFU/g, respectively. As Figure 3 shows, the samples at day 10 were already rotten, which mainly resulted in the destruction of plant tissues by the multiplication of spoilage microorganisms. Data obtained in our previous study showed that 405 nm light treatment effectively inactivates bacteria from *P. reinekei* and *P. palleroniana* [10]. These results urged us to further investigate the photodynamic inactivation effect and the actual preservation effect of Chl-based PDI. In this experiment, data showed that the populations of colonies in the light group and Chl+light group were significantly controlled ($p < 0.05$). The populations of colonies in the Chl and Chl+light groups decreased by 13.91% and 74.23%, respectively, compared to the control group by day 10. Compared to light sterilization alone, 70.07% more bacteria were inactivated with the addition of Chl, which significantly improved the efficiency of sterilization by 405 nm light. These data are consistent with the studies by Buchovec et al., (2017) [22] and Luksiene et al., (2019) [17]. This overall reduction in microbial contamination by Chl-based PDI treatment extended the disease-free period of the samples by 4 days (Figures 2d and 3). Therefore, Chl-based PDI treatment, which greatly enhances the efficiency of visible light inhibition, could effectively inhibit the reproduction of microorganisms and control the populations of microorganisms in quantity.

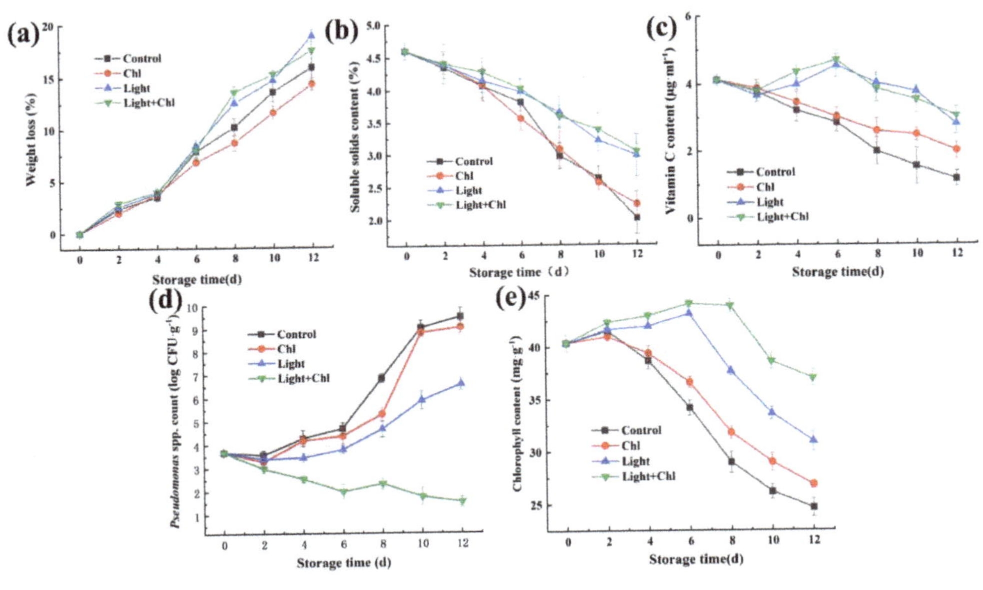

Figure 2. Changes in weight loss (**a**), soluble solids' content (**b**), chlorophyll content (**c**), vitamin C content (**d**), and *Pseudomonas* spp. count (**e**) of samples with different treatments. Error bars represent the mean ± SE ($n = 3$).

Figure 3. Change of samples in different groups during storage.

3.2. *Anti-Biofilm Effects of Chl-Based PDI against Inoculated Spoilage Microorganisms*

Biofilms contribute to the persistence of microbial contamination by shielding pathogens and spoilage bacteria from environmental stresses, acting as hot spots for horizontal gene transfer (HGT) of virulence genes, transforming previously benign strains into pathogens [33], and providing niches for antimicrobial-resistant mutagenic activities [34]. Thus, biofilms pose great potential hazards for food safety and human health, which prompted the development of the PDI technique. The anti-biofilm effect of the Chl-based PDI treatment was observed by using CLSM (Figure 4). On day 0, no biofilm was produced. However, at day 6, a small amount of biofilm appeared in both the Chl and control groups, while little biofilm was observed in the light and Chl+light groups (Figure 4—Day 6). Figure 5a shows the change of OD_{600} using crystal violet staining, which quantified the biofilm content. On day 6, compared to the control group, the amount of biofilm was reduced by 70.27% and 94.59% in the light and Chl+light groups, respectively ($p < 0.05$). On day 12, although the OD_{600} of the control group reached 1.03, the amount of biofilm of the light and Chl + light groups decreased by 55.34% and 88.35%, respectively ($p < 0.05$). This indicated that Chl-based PDI or light treatment effectively inhibited biofilm production during storage. The same results were found in studies on chili peppers [35], whole milk [36], salmon [37], and citrus fruits [38].

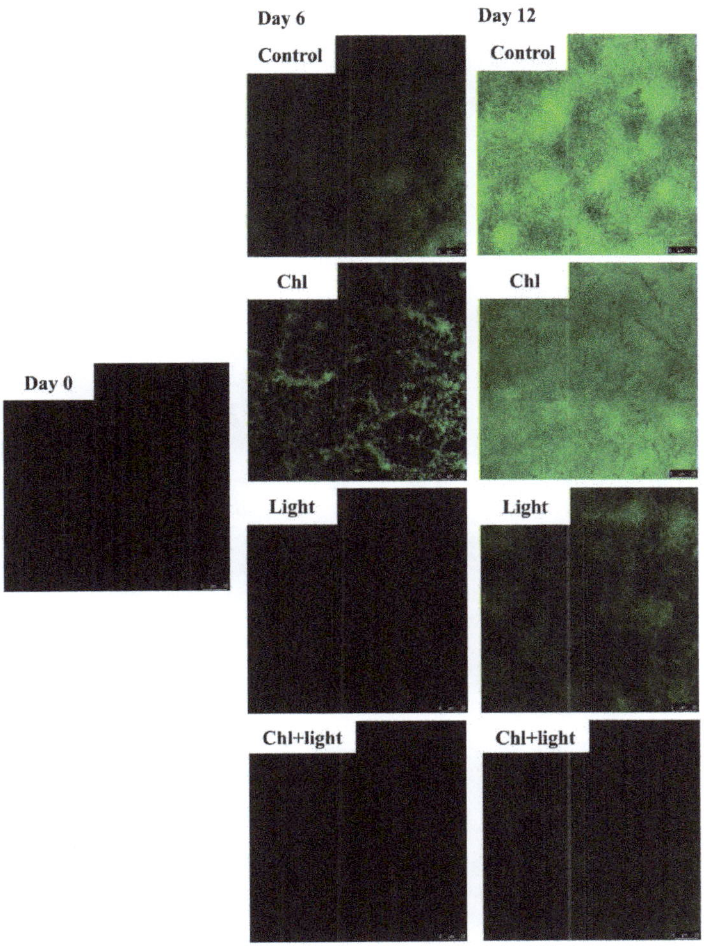

Figure 4. The figures of biofilms' formation on samples by using CLSM.

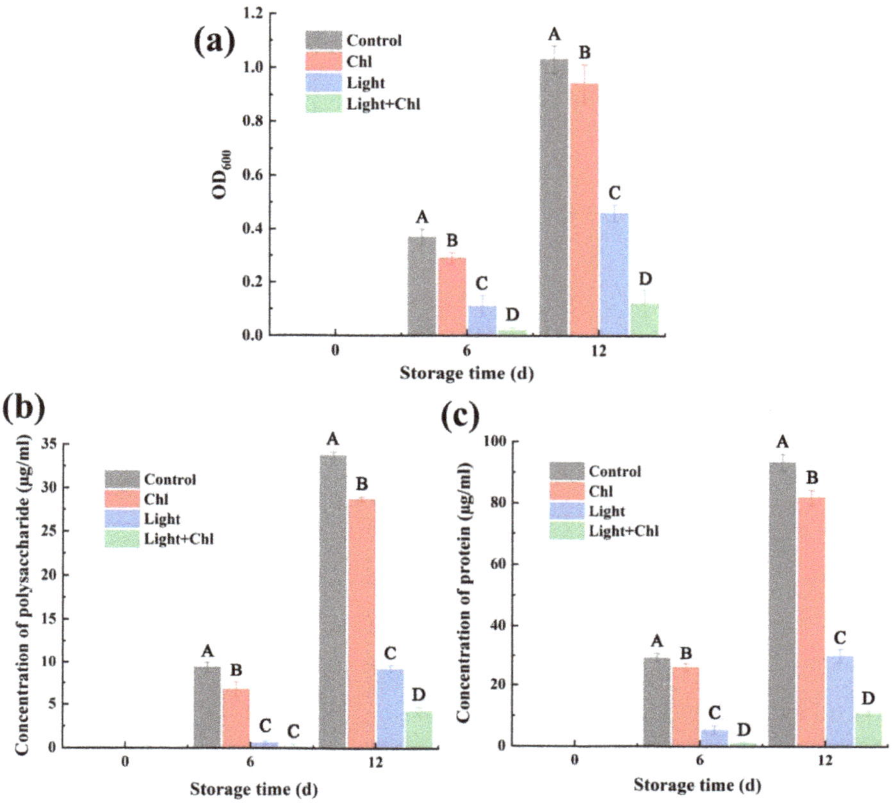

Figure 5. The biofilms' formation on samples (**a**). The effect of Chl-based PDI on extracellular polysaccharides' concentration (**b**) and extracellular protein concentration (**c**). A, B, C, and D mean significantly different in groups ($p < 0.05$). Error bars represent the mean ± SE ($n = 3$).

As shown in Figure 5b, the concentration of extracellular polysaccharides decreased by 72.84% and 87.41% after light or Chl+light treatment for 12 days, respectively ($p < 0.05$), while the concentration of extracellular protein decreased by 67.98% and 88.46%, respectively ($p < 0.05$, Figure 5c). Extracellular polymeric substance (EPS) consists mainly of substances such as extracellular proteins and extracellular polysaccharides. However, EPS secretion contributes to bacterial colonization or proliferation [39]. Obviously, Chl-based PDI significantly inhibited the secretion of extracellular polysaccharides and extracellular proteins, and the scavenging effect was significantly higher. These speculations were further validated by CLSM observations, which elucidated the effect of Chl-based PDI on the removal of biofilm. The infection of spoilage bacteria is a complex process involving factors such as colonization, proliferation, and biofilm formation. Colonization is achieved by specific appendages or by the production of EPS. In the present experiment, Chl is based on the type II PDI mechanism, through the generation of singly linear oxygen by coupling the excited state photosensitizer to the ground state oxygen, thereby killing cells [40]. In addition, the Chl-based PDI disrupted biofilms by disrupting biofilm structure and subsequent dispersion/reduction in attached microorganisms as well as by disrupting extracellular polymeric substance (EPS), including extracellular proteins and polysaccharides (Figure 6). This opens up the application of Chl-based PDI in addressing food safety associated with bacterial biofilms.

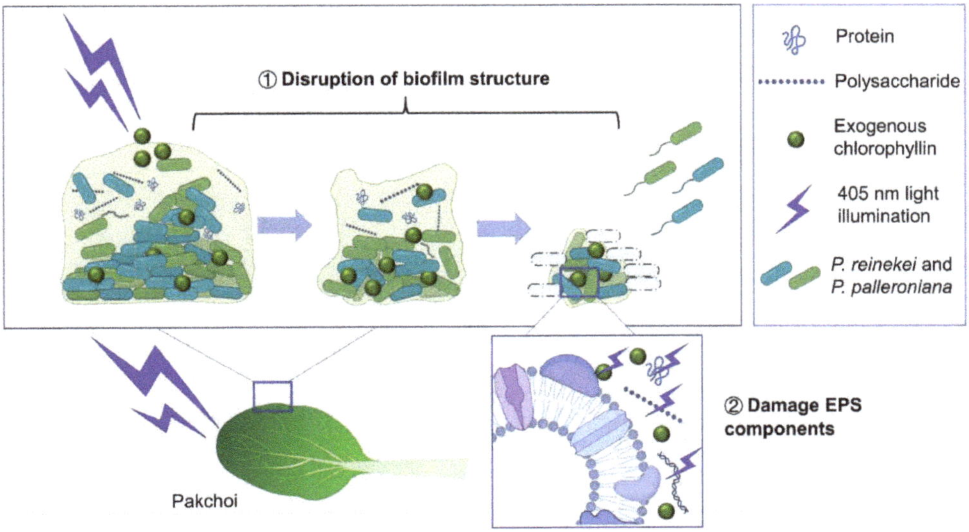

Figure 6. Schematic illustration of biofilm destruction by Chl-based PDI. Chl-based PDI disrupted biofilms by disrupting biofilm structure and subsequent dispersion/reduction in attached microorganisms as well as by disrupting extracellular polymeric substance (EPS), including extracellular proteins and polysaccharides.

3.3. Changes in Weight Loss and Water Distribution

During the storage stage, the increase in weight loss was caused by the water loss [41]. Since light causes the opening of the stomata of the leaves, light treatment might cause an increase in weight loss during storage [42]. We tested to learn if the light dose used in the present study negatively affected the samples. Figure 2a reveals that there was a gradual increase in weight loss of all groups during storage. From day 0 to day 12 of storage, the differences in weight loss were insignificant for all groups. However, at the beginning of day 8, the weight loss of the light and Chl+light groups was slightly higher than that of the control and Chl groups. This indicated that the dose of 22.27 J/cm^2 per day did not have a significant negative effect on the sample's weight loss.

Low-field NMR analysis helped us to better understand the moisture composition in the samples. Figure 7 shows the day 0 and day 12 water distribution of different samples. There were three peaks, representing the three relaxation components, called T21 (<2 ms), T22 (2–20 ms), and T23 (>1000 ms), which represent bound water, immobilized water, and free water, respectively (Figure 7) [43]. After 12 days of storage, the light and Chl+light treated groups' T23 peak decreased, which indicated that part of the free water had been lost. This was mainly caused by the transpiration and respiration of the leaves. When the steam pressure differed between the water in the blade and that generated outside, the water would dissipate through the pores [44]. The opening of the stomata accelerates this. Meanwhile, both the light and Chl+light groups had higher peak T22 and T21 than the other two groups, which may be due to the 405 nm light triggering the active internal biochemical reactions and accumulation of carbohydrates and sugars of the leaves [45]. Furthermore, the external environment such as the bacterial reproduction and low temperature will cause the adverse reactions, which lead to the participation of a large amount of water. Therefore, we speculated that the higher peak of T22 and T21 than the other two groups might have been due to the light and Chl-based photodynamic treatment alleviating the environmental stress during vegetable storage.

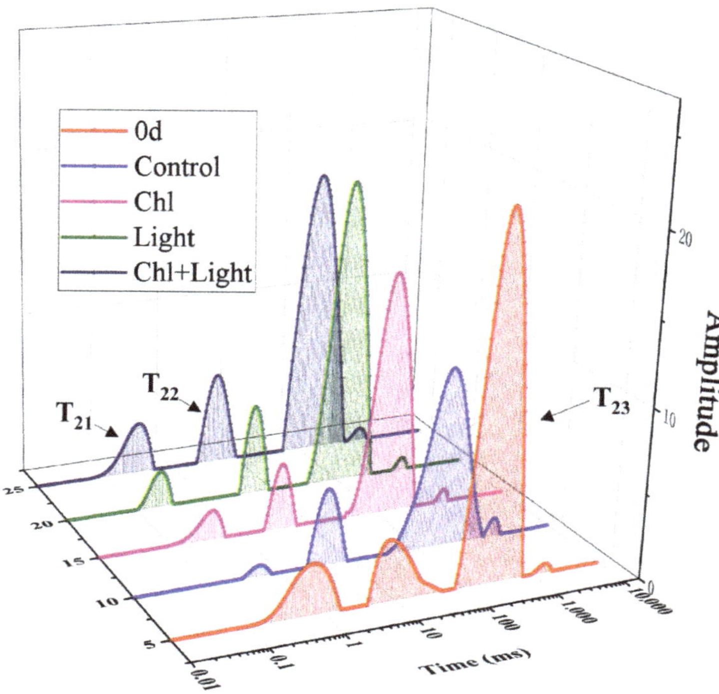

Figure 7. Changes in water distribution of samples with different treatments.

3.4. Changes in Total Soluble Solids (TSS) Content and Ascorbic Acid (AsA) Content

TSS in leafy vegetables includes soluble sugar, vitamins, and other water-soluble substances [42]. The initial TSS of the samples was 4.60 ± 0.13. Figure 2b shows that there was a gradual decrease in the content of TSS. However, the Chl and control groups had a sharply decreased TSS from day 8, which was significantly lower than the other groups ($p < 0.05$). The high TSS depletion observed in the Chl and control groups might have been caused by microbial spoilage or a higher energy consumption [46], while the other two groups, which were treated with light, showed a smaller decrease in TSS content. This was due to the activation of photosynthesis by a light note and the degradation of insoluble starch and other substances in the plants into soluble sugars, increasing the content of soluble sugars and other substances in the plants and supplementing the energy consumption of sugars and other substances in the leafy vegetables after harvest, thus maintaining the stability of the soluble solids' content. The decrease in TSS content of vegetables during refrigeration is mainly due to the depletion of energy metabolism, while microbial colonization accelerated the decrease in TSS. It has been suggested that, compared with the effect on vegetable tissue respiratory stimulation, light treatment has a greater effect on microbial inactivation [46]. Furthermore, TSS can help vegetables overcome low-temperature stress and improve freshness quality [47].

AsA is an important antioxidant substance in vegetables; it also affected the degree of vegetable browning. The initial AsA content of the samples was 4.13 ± 0.17 $\mu g \cdot mL^{-1}$. However, Figure 2d shows that the AsA content of different groups showed different trends. The ASA levels in both the light and Chl+light groups showed a trend of increasing and then decreasing, and reached a peak on day 6 (4.53 ± 0.33 and 4.69 ± 0.28 $\mu g \cdot mL^{-1}$, respectively). Since the ASA content of the control and Chl groups gradually decreased, it was significantly lower than that of other two groups on day 6 (2.84 ± 0.28 and 3.01 ± 0.28 $\mu g \cdot mL^{-1}$, respectively). The reason for this difference was the accumulation induced by light-activated respiration and photosynthesis [48]. Witkowska et al., (2013) proved that

AsA content and soluble carbohydrate content in fresh-cut lettuce through light treatment during storage also improved the appearance of vegetables and extended their shelf life [49]. However, as the storage time went on, the rate of the generated AsA could not keep up with the accumulation rate of oxidized substances caused by aging [50]. Therefore, the content of the light and Chl+light groups began to gradually decreased from day 6 and finally went down to 2.78 ± 0.31 and 3.03 ± 0.27 $\mu g \cdot mL^{-1}$ on day 12, respectively, still significantly higher than that of the control and Chl groups (1.14 ± 0.23 and 1.98 ± 0.25 $\mu g \cdot mL^{-1}$, respectively, $p < 0.05$). These results are consistent with the finding of Tao et al., (2019) [46] and Witkowska et al., (2013) [48], that light causes a trend of increasing and then gradually decreasing AsA content during storage.

3.5. Changes in Chlorophyll Content, Color, and Organoleptic Properties

Chlorophyll content was an important index to evaluate the senescence and commercial value of leafy vegetables. Chlorophyll, a tetrapyrrole compound containing magnesium ions, easily replaced magnesium ions with hydrogen ions in dark conditions during storage, causing plant leaves to brown and yellow [45]. The initial content of chlorophyll was 40.36 ± 0.85 $mg \cdot g^{-1}$. The chlorophyll content showed a trend of increasing and then decreasing. However, the chlorophyll content of the control group and the Chl-only group increased up to day 2, 41.53 ± 0.67, 40.98 ± 0.4 $mg \cdot g^{-1}$, respectively. After that, the chlorophyll content gradually decreased and finally went down to 24.62 ± 0.84 and 26.77 ± 0.35 $mg \cdot g^{-1}$ at day 12, respectively. It was shown that the decrease in the chlorophyll content during storage might have been due to the rupture of the intracellular membrane, resulting in the degradation of chlorophyll by the action of chlorophyllase. However, the light treatment delayed the occurrence of the turning point of the change of chlorophyll content, in which the content of the light and Chl+light groups increased until day 6 (43.17 ± 0.15, 44.13 ± 0.33 $mg \cdot g^{-1}$, respectively). The chlorophyll content then gradually decreased until day 12, when it went down to 30.92 ± 0.93, 36.92 ± 0.8 $mg \cdot g^{-1}$, respectively. From day 4 until the end of storage, the chlorophyll content of the light and Chl+light groups was significantly higher than that of the other two groups ($p < 0.05$). Chlorophyll loss in leafy vegetables, which leads to leaf yellowing or browning, usually implies senescence. However, the low intensity of the LED light increased chlorophyll production and delayed senescence during storage [51–54]. Therefore, the low dose of light treatment in the present experiment effectively enhanced the chlorophyll content.

Color is one of the main sensory characteristics of fresh vegetables and determines the consumer acceptance of agricultural products. Figure 3 shows photos of different samples during storage. Table 1 shows the changes in L* of the samples. The larger L* value, the brighter the color was; the opposite represents a darker color. The L* of the control group showed a trend of increasing and then decreasing. Combined with the results from Figure 3, it can be seen that the fresh samples showed a dark green color and presented a lower L* value; as the storage time went on and the degradation of chlorophyll occurred, the color of the samples gradually became lighter and the L* value gradually increased. At the end of storage, the L* value decreased again due to the severe browning of the control group, which showed a yellowish-brown color. The same trend occurred in the Chl group that lacked light treatment. However, the L* value of the light and Chl+light groups had a smaller fluctuation range, which indicated that the light treatment could keep the color of the samples in a more stable range.

Tables 2 and 3 show the a* and b* values of each group during storage, respectively. The a* represents a color bias towards red (+) or green (−), and b* represents a color bias towards yellow (+) or blue (−). The initial value of a* was −17.99 ± 2.76, and the fresh samples appeared dark green. However, the a* values of the samples from the control group gradually increased during the storage, which indicated that the color of the samples increasingly took on a yellow hue. From day 8, the a* value of the control group (−7.85 ± 3.78) was significantly higher than that of the previous 6 days and higher than that of the other three groups. This indicated that the samples in the control group started to yellow severely on day 8 and presented more severely than all the other groups. The b* value of the control group increased and then decreased, which was caused by the leaves turning yellow in the middle stage of storage and gradually getting brown under the action of microorganisms in the later stage. This is consistent with the results found in other studies, that microorganisms cause accelerated browning of leaves [6–8]. However, the yellowing of the leaves was better inhibited by the light and Chl+light groups. The light group showed significant changes only from day 10 ($p < 0.05$), while the Chl+light group also showed no significant changes in a* values during the storage. These indicated that light significantly inhibits yellowing or microbially induced browning of vegetables during storage. Similar results were found for LED treatment of fresh-cut lettuce [46], purple kale [55], and baby mustard [56] during storage.

To determine the effect of the Chl-based PDI treatment on the samples, a panel of 23 judges rated the samples according to their senses. According to the results obtained (Figure 8), the Chl+light group achieved better sensory scores in terms of color, form, and smell. Additionally, the judging panel did not experience any off-flavors of the Chl-based PDI treatment samples. These results showed that Chl-based PDI had no negative effect on the samples. However, on the 10th day and 12th day, the control group was spoiled with off-odor and browning, which were regarded as unacceptable by the panelists. Combined with the form analysis, the low grade of organoleptic properties of the control group was mainly due to the peculiar smell, leaf yellowing, and wilting.

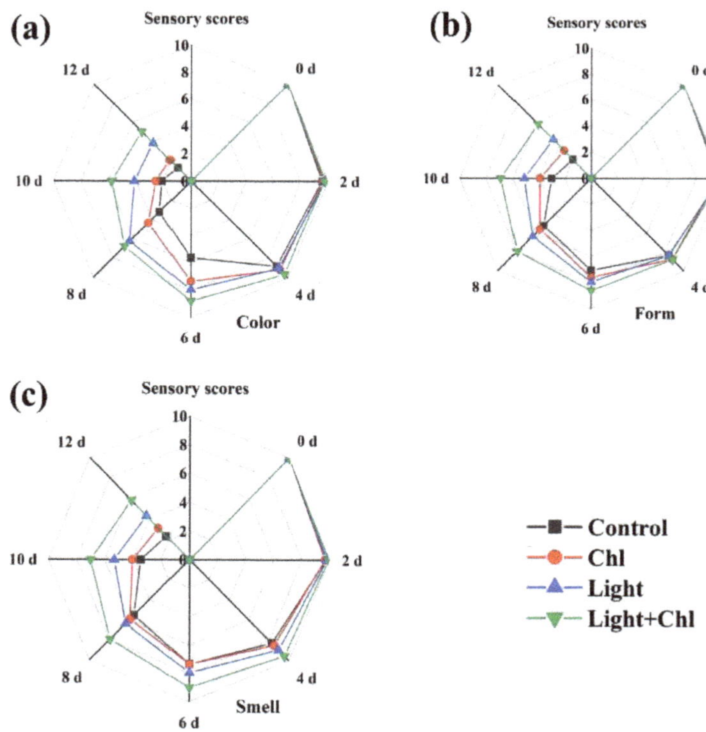

Figure 8. Changes in color (**a**), form (**b**) and smell (**c**) of organoleptic properties' results in different groups.

Table 1. L* value of samples during storage.

	0 d	2 d	4 d	6 d	8 d	10 d	12 d
Control	36.12 ± 2.55 ^{Ad}	40.69 ± 3.26 ^{Ad}	37.73 ± 2.78 ^{Bd}	49.67 ± 2.55 ^{Ac}	69.9 ± 3.95 ^{Aa}	61.49 ± 3.96 ^{Ab}	54.21 ± 16.98 ^{Abc}
Light	36.12 ± 2.55 ^{Ab}	41.27 ± 1.93 ^{Aab}	42.7 ± 1.64 ^{Aab}	39.36 ± 3.17 ^{Cab}	36.53 ± 3.57 ^{Db}	44.64 ± 5.43 ^{Ba}	42.62 ± 7.01 ^{Aab}
Chl	36.12 ± 2.55 ^{Ae}	42.51 ± 1.49 ^{Abc}	38.02 ± 2.23 ^{Bde}	44.86 ± 1.73 ^{Bb}	50.15 ± 3.65 ^{Ba}	40.53 ± 3.76 ^{Bcd}	50.38 ± 3.71 ^{Aa}
Chl+Light	36.12 ± 2.55 ^{Acd}	38.92 ± 3.08 ^{Ac}	32.99 ± 1.39 ^{Cd}	34.12 ± 3.17 ^{Dcd}	43.67 ± 3.79 ^{Cb}	44.59 ± 5.13 ^{Bb}	52.38 ± 4.00 ^{Aa}

$^{A, B, C, D}$ Means in the same column with different superscripts are significantly different ($p < 0.05$). $^{a, b, c, d}$ Means in the same row with different superscripts are significantly different ($p < 0.05$).

Table 2. The a* value of samples during storage.

	0 d	2 d	4 d	6 d	8 d	10 d	12 d
Control	−17.99 ± 2.76 ^{Ac}	−19.69 ± 2.47 ^{Ac}	−18.87 ± 0.64 ^{Bc}	−17.69 ± 1.75 ^{Ac}	−7.85 ± 3.78 ^{Ab}	4.78 ± 2.54 ^{Aa}	7.73 ± 2.77 ^{Aa}
Light	−17.99 ± 2.76 ^{Ac}	−19.31 ± 0.99 ^{Ac}	−19.01 ± 0.78 ^{Bc}	−18.05 ± 1.00 ^{Ac}	−15.77 ± 1.17 ^{Bc}	−10.76 ± 4.05 ^{Cb}	−7.56 ± 3.52 ^{Ca}
Chl	−17.99 ± 2.76 ^{Acd}	−18.27 ± 1.10 ^{Acd}	−18.69 ± 0.88 ^{Bcd}	−19.91 ± 0.94 ^{Bd}	−15.10 ± 1.18 ^{Bc}	−4.72 ± 4.64 ^{Bb}	3.78 ± 2.34 ^{Ba}
Chl+Light	−17.99 ± 2.76 ^{Aa}	−18.81 ± 1.85 ^{Aa}	−17.02 ± 0.57 ^{Aa}	−17.17 ± 1.42 ^{Aa}	−18.86 ± 0.54 ^{Ca}	−17.3 ± 1.38 ^{Da}	−16.79 ± 2.20 ^{Da}

$^{A, B, C, D}$ Means in the same column with different superscripts are significantly different ($p < 0.05$). $^{a, b, c, d}$ Means in the same row with different superscripts are significantly different ($p < 0.05$).

Table 3. The b* value of samples during storage.

	0 d	2 d	4 d	6 d	8 d	10 d	12 d
Control	34.37 ± 1.51 ^{Ac}	34.57 ± 2.95 ^{Ac}	35.27 ± 2.20 ^{Bc}	44.61 ± 2.69 ^{Ab}	65.61 ± 4.07 ^{Aa}	50.60 ± 2.87 ^{Ab}	47.76 ± 15.00 ^{Ab}
Light	34.37 ± 1.51 ^{Aa}	35.39 ± 3.77 ^{Aa}	41.31 ± 2.08 ^{Aa}	34.17 ± 1.50 ^{Ca}	35.28 ± 3.02 ^{Ca}	39.79 ± 8.67 ^{Ba}	34.65 ± 5.32 ^{Ba}
Chl	34.37 ± 1.51 ^{Ab}	35.23 ± 3.68 ^{Ab}	37.26 ± 1.40 ^{Bb}	38.05 ± 2.69 ^{Bb}	44.62 ± 5.00 ^{Ba}	34.87 ± 4.67 ^{Bb}	37.88 ± 3.63 ^{ABb}
Chl+Light	34.37 ± 1.51 ^{Ac}	34.54 ± 2.20 ^{Ac}	32.49 ± 1.78 ^{Cc}	30.77 ± 1.63 ^{Dc}	39.22 ± 1.45 ^{Cb}	42.11 ± 4.13 ^{Bb}	48.56 ± 3.53 ^{Aa}

$^{A, B, C, D}$ Means in the same column with different superscripts are significantly different ($p < 0.05$). $^{a, b, c, d}$ Means in the same row with different superscripts are significantly different ($p < 0.05$).

3.6. Changes in Antioxidant Enzyme Activity

During storage, the content of reactive oxygen species (ROS) increases with the aging of leaves [57]. Meanwhile, the accumulation of ROS, such as H_2O_2 and O^{2-} causes oxidation stress or oxidation damage to cells [58]. SOD, the main antioxidant enzyme in vegetables, plays an important part in eliminating excessive ROS in vegetables [59]. Therefore, it is necessary to investigate whether changes in antioxidant activity occurred after photosensitization treatment. Figure 9 shows the changes in POD and SOD activity. The POD activity of each group gradually increased during storage (Figure 9a). From day 6, the activity of the control or Chl groups was significantly higher than that of the two groups with light treatment ($p < 0.05$). This indicated that light effectively inhibited the increase in POD activity. The increase in POD activity often leads to the browning of vegetables [60]. Therefore, light treatment had the potential to reduce vegetable browning by inhibiting POD activity. The SOD content showed a trend of increasing and then decreasing (Figure 9b). From day 4 onwards, the two groups with light treatment showed significantly higher SOD activity than the other two groups. In fact, plant cells usually tightly control ROS levels by producing or activating antioxidant enzymes. This suggested that light treatment or photodynamic treatment helped to enhance antioxidant enzyme activity or reduce ROS levels in vegetables during storage. Similar results were found in pakchoi [45] and fresh-cut celery [50].

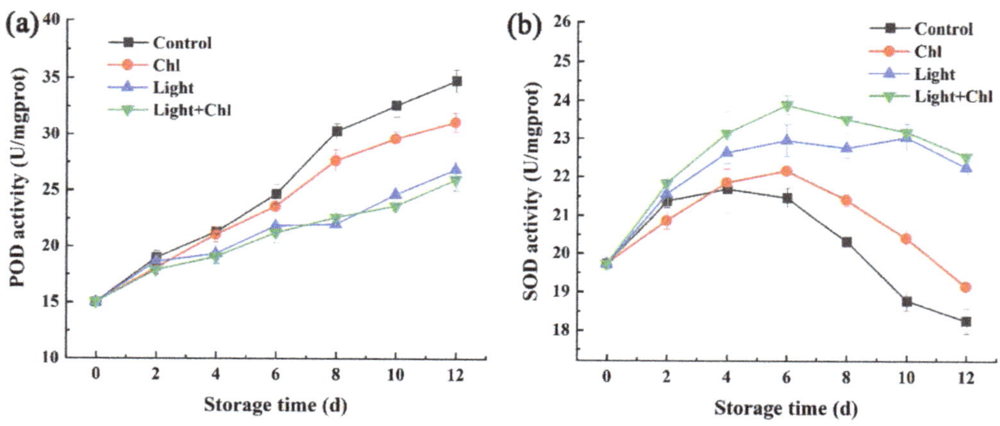

Figure 9. Changes in POD (**a**) and SOD (**b**) activity of samples. Error bars represent the mean ± SE ($n = 3$).

4. Conclusions

In conclusion, Chl-based PDI effectively improved pakchoi quality at postharvest and inhibited the formation of biofilm. The results showed that the Chl-based PDI increased the visual quality and the content of chlorophyll, VC, total soluble solids, and SOD activity and decreased the occurrence of leaf yellowing, populations of inoculum bacterial, and POD activity. Additionally, the Chl-based PDI reduced the amount of biofilm and inhibited the production of bacterial extracellular polysaccharides and extracellular proteins. The experimental data supported the view that a new generation of photoactive chlorophyllin with 405 nm light could effectively preserve fresh-cut pakchoi in a non-thermal way, which is in line with the requirements of clean and green technology. Thus, PDI has the potential to be used as an antimicrobial tool for treating fresh-cut fruits and vegetables or other food products. However, it also has limitations; for example, it cannot effectively sterilize areas that cannot be reached by light. The feasibility of its application in practical production needs to be investigated in the future.

Author Contributions: Conceptualization, Y.Z., Z.D., C.S. and J.X.; methodology, Y.Z.; software, Y.Z.; validation, Y.Z. and J.X.; investigation, Y.Z.; resources, J.X.; writing—original draft preparation, Y.Z.; writing—review and editing, Z.D. and J.X.; visualization, Y.Z.; supervision, J.X.; project administration, J.X.; funding acquisition, J.X. All authors have read and agreed to the published version of the manuscript.

Funding: This research was funded by the Shanghai Green Leaf Vegetable Industry Technology System Project (20DZ2292200) and the Shanghai Professional Technology Service Platform on Cold Chain Equipment Performance and Energy Saving Evaluation (19DZ1207503).

Data Availability Statement: Data are contained within the article.

Conflicts of Interest: The authors declare no conflict of interest.

References

1. Mir, S.A.; Shah, M.A.; Mir, M.M.; Dar, B.; Greiner, R.; Roohinejad, S. Microbiological contamination of ready-to-eat vegetable salads in developing countries and potential solutions in the supply chain to control microbial pathogens. *Food Control* **2018**, *85*, 235–244. [CrossRef]
2. Snyder, A.B.; Worobo, R.W. The incidence and impact of microbial spoilage in the production of fruit and vegetable juices as reported by juice manufacturers. *Food Control* **2018**, *85*, 144–150. [CrossRef]
3. FAO. *The State of Food and Agriculture: Moving Forward on Food Loss and Waste Reduction*; FAO, Global Rome: Rome, Italy, 2019.
4. Wu, W.J.; Gao, H.Y.; Chen, H.J.; Fang, X.J.; Han, Q.; Zhong, Q.L. Combined effects of aqueous chlorine dioxide and ultrasonic treatments on shelf-life and nutritional quality of bok choy (*Brassica chinensis*). *LWT-Food Sci. Technol.* **2019**, *101*, 757–763. [CrossRef]
5. Federico, B.; Pinto, L.; Quintieri, L.; Canto, A.; Calabrese, N.; Caputo, L. Efficacy of lactoferricin B in controlling ready-to-eat vegetable spoilage caused by *Pseudomonas* spp. *Int. J. Food Microbiol.* **2015**, *215*, 179–186. [CrossRef] [PubMed]
6. Grogan, R.G. Varnish spot, destructive disease of lettuce in California caused by *Pseudomonas cichorii*. *Phytopathology* **1977**, *67*, 957–960. [CrossRef]
7. Lee, D.H.; Kim, J.-B.; Kim, M.; Roh, E.; Jung, K.; Choi, M.; Oh, C.; Choi, J.; Yun, J.; Heu, S. Microbiota on spoiled vegetables and their characterization. *J. Food Prot.* **2013**, *76*, 1350–1358. [CrossRef] [PubMed]
8. Liao, C.H. An extracellular pectate lyase is the pathogenicity factor of the soft-rotting *Bacterium Pseudomonas* viridiflava. *Mol. Plant-Microbe Interact.* **1988**, *1*, 199. [CrossRef]
9. Zhang, Y.; Xie, J. The effect of red and violet light emitting diode (LED) treatments on the postharvest quality and biodiversity of fresh-cut pakchoi (*Brassica rapa* L. Chinensis). *Food Sci. Technol. Int.* **2021**, *28*, 297–308. [CrossRef]
10. Zhang, Y.; Xie, J. Effects of 405 nm light-emitting diode treatment on microbial community on fresh-cut pakchoi and antimicrobial action against Pseudomonas reinekei and Pseudomonas palleroniana. *J. Food Saf.* **2021**, *41*, e12920. [CrossRef]
11. Morris, C.E.; Monier, J.-M.; Jacques, M.-A. A technique to quantify the population size and composition of the biofilm component in communities of bacteria in the phyllosphere. *Appl. Environ. Microbiol.* **1998**, *64*, 4789–4795. [CrossRef]
12. Cui, H.; Zhang, C.; Li, C.; Lin, L. Inhibition of Escherichia coli O157: H7 biofilm on vegetable surface by solid liposomes of clove oil. *LWT* **2020**, *117*, 108656. [CrossRef]
13. Artes-Hernandez, F.; Robles, P.A.; Gomez, P.A.; Tomas-Callejas, A.; Artes, F. Low UV-C illumination for keeping overall quality of fresh-cut watermelon. *Postharvest Biol. Technol.* **2010**, *55*, 114–120. [CrossRef]
14. Issazacharia, A.; Kamitani, Y.; Muhimbula, H.S.; Ndabikunze, B.K. A review of microbiological safety of fruits and vegetables and the introduction of electrolyzed water as an alternative to sodium hypochlorite solution. *Afr. J. Food Sci.* **2010**, *4*, 778–789.
15. Aponiene, K.; Paskeviciute, E.; Reklaitis, I.; Luksiene, Z. Reduction of microbial contamination of fruits and vegetables by hypericin-based photosensitization: Comparison with other emerging antimicrobial treatments. *J. Food Eng.* **2015**, *144*, 29–35. [CrossRef]
16. Luksiene, Z.; Paskeviciute, E. Novel approach to the microbial decontamination of strawberries: Chlorophyllin-based photosensitization. *J. Appl. Microbiol.* **2011**, *110*, 1274–1283. [CrossRef]
17. Luksiene, Z.; Buchovec, I. Impact of chlorophyllin-chitosan coating and visible light on the microbial contamination, shelf life, nutritional and visual quality of strawberries. *Innov. Food Sci. Emerg. Technol.* **2019**, *52*, 463–472. [CrossRef]
18. Tumolo, T.; Lanfer-Marquez, U.M. Copper chlorophyllin: A food colorant with bioactive properties? *Food Res. Int.* **2012**, *46*, 451–459. [CrossRef]
19. Luksiene, Z.; Buchovec, I.; Viskelis, P. Impact of high-power pulsed light on microbial contamination, health promoting components and shelf life of strawberries. *Food Technol. Biotechnol.* **2013**, *51*, 284–292.
20. Luksiene, Z.; Brovko, L. Antibacterial photosensitization-based treatment for food safety. *Food Eng. Rev.* **2013**, *5*, 185–199. [CrossRef]
21. Paskeviciute, E.; Zudyte, B.; Luksiene, Z. Towards better microbial safety of fresh produce: Chlorophyllin-based photosensitization for microbial control of foodborne pathogens on cherry tomatoes. *J. Photochem. Photobiol. B Biol.* **2018**, *182*, 130–136. [CrossRef]

22. Buchovec, I.; Lukseviciute, V.; Kokstaite, R.; Labeikyte, D.; Kaziukonyte, L.; Luksiene, Z. Inactivation of Gram (-) bacteria Salmonella enterica by chlorophyllin-based photosensitization: Mechanism of action and new strategies to enhance the inactivation efficiency. *J. Photochem. Photobiol. B Biol.* **2017**, *172*, 1–10. [CrossRef]
23. Cossu, M.; Ledda, L.; Cossu, A. Emerging trends in the photodynamic inactivation (PDI) applied to the food decontamination. *Food Res. Int.* **2021**, *144*, 110358. [CrossRef] [PubMed]
24. Chen, B.; Huang, J.; Li, H.; Zeng, Q.-H.; Wang, J.J.; Liu, H.; Pan, Y.; Zhao, Y. Eradication of planktonic Vibrio parahaemolyticus and its sessile biofilm by curcumin-mediated photodynamic inactivation. *Food Control* **2020**, *113*, 107181. [CrossRef]
25. Silva, A.F.; Borges, A.; Freitas, C.F.; Hioka, N.; Mikcha, J.M.G.; Simões, M. Antimicrobial photodynamic inactivation mediated by rose bengal and erythrosine is effective in the control of food-related bacteria in planktonic and biofilm states. *Molecules* **2018**, *23*, 2288. [CrossRef]
26. Bonifácio, D.; Martins, C.; David, B.; Lemos, C.; Neves, M.; Almeida, A.; Pinto, D.; Faustino, M.; Cunha, Â. Photodynamic inactivation of Listeria innocua biofilms with food-grade photosensitizers: A curcumin-rich extract of Curcuma longa vs. commercial curcumin. *J. Appl. Microbiol.* **2018**, *125*, 282–294. [CrossRef]
27. Zhang, Y.; Ding, Z.; Xie, J. Metabolic effects of violet light on spoilage bacteria from fresh-cut pakchoi during postharvest stage. *Plants* **2022**, *11*, 267. [CrossRef]
28. Hasperué, J.H.; Guardianelli, L.; Rodoni, L.M.; Chaves, A.R.; Martínez, G.A. Continuous white–blue LED light exposition delays postharvest senescence of broccoli. *LWT Food Sci. Technol.* **2016**, *65*, 495–502. [CrossRef]
29. Wang, Q.; Ding, T.; Zuo, J.; Gao, L.; Fan, L. Amelioration of postharvest chilling injury in sweet pepper by glycine betaine. *Postharvest Biol. Technol.* **2016**, *112*, 114–120. [CrossRef]
30. He, J.; Ren, Y.; Chen, X.; Chen, H. Protective roles of nitric oxide on seed germination and seedling growth of rice (*Oryza sativa* L.) under cadmium stress. *Ecotoxicol. Environ. Saf.* **2014**, *108*, 114–119. [CrossRef]
31. Zhang, W.-H.; Wu, J.; Weng, L.; Zhang, H.; Zhang, J.; Wu, A. An improved phenol-sulfuric acid method for the determination of carbohydrates in the presence of persulfate. *Carbohydr. Polym.* **2020**, *227*, 115332. [CrossRef]
32. Yan, J.; Xie, J. Comparative proteome analysis of Shewanella putrefaciens WS13 mature biofilm under cold stress. *Front. Microbiol.* **2020**, *11*, 1225. [CrossRef]
33. Stalder, T.; Top, E. Plasmid transfer in biofilms: A perspective on limitations and opportunities. *NPJ Biofilms Microbiomes* **2016**, *2*, 16022. [CrossRef]
34. Flemming, H.-C.; Wingender, J.; Szewzyk, U.; Steinberg, P.; Rice, S.A.; Kjelleberg, S. Biofilms: An emergent form of bacterial life. *Nat. Rev. Microbiol.* **2016**, *14*, 563–575. [CrossRef]
35. Tortik, N.; Spaeth, A.; Plaetzer, K. Photodynamic decontamination of foodstuff from Staphylococcus aureus based on novel formulations of curcumin. *Photochem. Photobiol. Sci.* **2014**, *13*, 1402–1409. [CrossRef]
36. Chorianopoulos, N.; Tsoukleris, D.; Panagou, E.; Falaras, P.; Nychas, G.-J. Use of titanium dioxide (TiO_2) photocatalysts as alternative means for Listeria monocytogenes biofilm disinfection in food processing. *Food Microbiol.* **2011**, *28*, 164–170. [CrossRef]
37. Li, X.; Kim, M.-J.; Bang, W.-S.; Yuk, H.-G. Anti-biofilm effect of 405-nm LEDs against Listeria monocytogenes in simulated ready-to-eat fresh salmon storage conditions. *Food Control* **2018**, *84*, 513–521. [CrossRef]
38. Yu, S.-M.; Lee, Y.H. Effect of light quality on Bacillus amyloliquefaciens JBC36 and its biocontrol efficacy. *Biol. Control.* **2013**, *64*, 203–210. [CrossRef]
39. Sun, Y.; Ma, Y.; Guan, H.; Liang, H.; Zhao, X.; Wang, D. Adhesion mechanism and biofilm formation of Escherichia coli O157: H7 in infected cucumber (*Cucumis sativus* L.). *Food Microbiol.* **2022**, *105*, 103885. [CrossRef]
40. Sheng, L.; Li, X.; Wang, L. Photodynamic inactivation in food systems: A review of its application, mechanisms, and future perspective. *Trends Food Sci. Technol.* **2022**, *124*, 167–181. [CrossRef]
41. Wang, Y.; Yi, Y.; Zhang, Y.; Guo, Z.; Liping, B.; Zhang, J.; Hou, L. Physiological response of winter wheat seedlings to water stress. *Jiangsu Agric. Sci.* **2013**, *41*, 75–76.
42. Hasperué, J.H.; Chaves, A.R.; Martínez, G.A. End of day harvest delays postharvest senescence of broccoli florets. *Postharvest Biol. Technol.* **2011**, *59*, 64–70. [CrossRef]
43. Mothibe, K.J.; Zhang, M.; Mujumdar, A.S.; Wang, Y.C.; Cheng, X.F. Effects of ultrasound and microwave pretreatments of apple before spouted bed drying on rate of dehydration and physical properties. *Dry. Technol.* **2014**, *32*, 1848–1856. [CrossRef]
44. Bartz, J.A.; Brecht, J.K. *Postharvest Physiology and Pathology of Vegetables*; CRC Press: Boca Raton, FL, USA, 2002.
45. Zhou, F.; Zuo, J.; Xu, D.; Gao, L.; Wang, Q.; Jiang, A. Low intensity white light-emitting diodes (LED) application to delay senescence and maintain quality of postharvest pakchoi (*Brassica campestris* L. ssp. *chinensis* (L.) *Makino* var. communis Tsen et Lee). *Sci. Hortic.* **2020**, *262*, 109060. [CrossRef]
46. Tao, T.; Ding, C.; Han, N.; Cui, Y.; Liu, X.; Zhang, C. Evaluation of pulsed light for inactivation of foodborne pathogens on fresh-cut lettuce: Effects on quality attributes during storage. *Food Packag. Shelf Life* **2019**, *21*, 100358. [CrossRef]
47. Burdon, J.; Lallu, N.; Francis, K.; Boldingh, H. The susceptibility of kiwifruit to low temperature breakdown is associated with pre-harvest temperatures and at-harvest soluble solids content. *Postharvest Biol. Technol.* **2007**, *43*, 283–290. [CrossRef]
48. Ntagkas, N.; Woltering, E.J.; Marcelis, L.F. Light regulates ascorbate in plants: An integrated view on physiology and biochemistry. *Environ. Exp. Bot.* **2018**, *147*, 271–280. [CrossRef]
49. Witkowska, I.M. Factors Affecting the Postharvest Performance of Fresh-Cut Lettuce. Ph.D. Thesis, Wageningen University, Wageningen, The Netherlands, 2013.

50. Zhan, L.J.; Hu, J.Q.; Pang, L.Y.; Li, Y.; Shao, J.F. Effects of light exposure on chlorophyll, sugars and vitamin C content of fresh-cut celery (*Apium graveolens* var. dulce) petioles. *Int. J. Food Sci. Technol.* **2014**, *49*, 347–353. [CrossRef]
51. Nassarawa, S.S.; Abdelshafy, A.M.; Xu, Y.; Li, L.; Luo, Z. Effect of light-emitting diodes (LEDs) on the quality of fruits and vegetables during postharvest period: A review. *Food Bioprocess Technol.* **2021**, *14*, 388–414. [CrossRef]
52. Braidot, E.; Petrussa, E.; Peresson, C.; Patui, S.; Bertolini, A.; Tubaro, F.; Wählby, U.; Coan, M.; Vianello, A.; Zancani, M. Low-intensity light cycles improve the quality of lamb's lettuce (*Valerianella olitoria* [L.] Pollich) during storage at low temperature. *Postharvest Biol. Technol.* **2014**, *90*, 15–23. [CrossRef]
53. Zhan, L.; Hu, J.; Li, Y.; Pang, L. Combination of light exposure and low temperature in preserving quality and extending shelf-life of fresh-cut broccoli (*Brassica oleracea* L.). *Postharvest Biol. Technol.* **2012**, *72*, 76–81. [CrossRef]
54. Liu, J.D.; Goodspeed, D.; Sheng, Z.; Li, B.; Yang, Y.; Kliebenstein, D.J.; Braam, J. Keeping the rhythm: Light/dark cycles during postharvest storage preserve the tissue integrity and nutritional content of leafy plants. *BMC Plant Biol.* **2015**, *15*, 92. [CrossRef]
55. Bárcena, A.; Martínez, G.; Costa, L. Low intensity light treatment improves purple kale (*Brassica oleracea* var. sabellica) postharvest preservation at room temperature. *Heliyon* **2019**, *5*, e02467. [CrossRef]
56. Sun, B.; Di, H.; Zhang, J.; Xia, P.; Huang, W.; Jian, Y.; Zhang, C.; Zhang, F. Effect of light on sensory quality, health-promoting phytochemicals and antioxidant capacity in post-harvest baby mustard. *Food Chem.* **2021**, *339*, 128057. [CrossRef]
57. Zimmermann, P.; Zentgraf, U. The correlation between oxidative stress and leaf senescence during plant development. *Cell. Mol. Biol. Lett.* **2005**, *10*, 515–534.
58. Ren, Y.F.; Wang, W.; He, J.Y.; Zhang, L.Y.; Wei, Y.J.; Yang, M. Nitric oxide alleviates salt stress in seed germination and early seedling growth of pakchoi (*Brassica chinensis* L.) by enhancing physiological and biochemical parameters. *Ecotoxicol. Environ. Saf.* **2020**, *187*, 109785. [CrossRef]
59. Lin, K.H.R.; Weng, C.C.; Lo, H.F.; Chen, J.T. Study of the root antioxidative system of tomatoes and eggplants under waterlogged conditions. *Plant Sci.* **2004**, *167*, 355–365. [CrossRef]
60. Pristijono, P.; Wills, R.B.H.; Golding, J.B. Inhibition of browning on the surface of apple slices by short term exposure to nitric oxide (NO) gas. *Postharvest Biol. Technol.* **2006**, *42*, 256–259. [CrossRef]

foods

Article

Transformation of Inferior Tomato into Preservative: Fermentation by Multi-Bacteriocin Producing *Lactobacillus paracasei* WX322

Rong Zhu [1], Xiaoqing Liu [1], Xiaofen Li [1], Kaifang Zeng [1,2] and Lanhua Yi [1,2,*]

[1] College of Food Science, Southwest University, Chongqing 400715, China; zr852963@email.swu.edu.cn (R.Z.); lyj594lxq@email.swu.edu.cn (X.L.); lixiaofen222@email.swu.edu.cn (X.L.); zengkaifang@hotmail.com (K.Z.)
[2] Research Center of Food Storage & Logistics, Southwest University, Chongqing 400715, China
* Correspondence: yilanhua@swu.edu.cn; Tel./Fax: +86-23-68250374

Abstract: Loss and waste of postharvest vegetables are the main challenges facing the world's vegetable supply. In this study, an innovative method of value-added transformation was provided: production of bacteriocin from vegetable waste, and then its application to preservation of vegetables. Antibacterial activity to soft rot pathogen *Pectobacterium cartovorum* (*Pcb* BZA12) indicated that tomato performed best in the nutrition supply for bacteriocin production among 12 tested vegetables. Moreover, the antibacterial activity was from *Lactobacillus paracasei* WX322, not components of vegetables. During a fermentation period of 10 days in tomato juice, *L. paracasei* WX322 grew well and antibacterial activity reached the maximum on the tenth day. Thermostability and proteinase sensitivity of the bacteriocin from tomato juice were the same with that from Man-Rogosa-Sharpe broth. Scanning electron microscope images indicated that the bacteriocin from tomato juice caused great damage to *Pcb* BZA12. At the same time, the bacteriocin from tomato juice significantly reduced the rotten rate of Chinese cabbage from 100% ± 0% to 20% ± 8.16% on the third day during storage. The rotten rate decrease of cucumber, tomato, and green bean was 100% ± 0% to 0% ± 0%, 70% ± 14.14% to 13.33% ± 9.43%, and 76.67% ± 4.71% to 26.67% ± 4.71%, respectively. Bacteriocin treatment did not reduce the rotten rate of balsam pear, but alleviated its symptoms.

Keywords: tomato; bacteriocin; soft rot; biocontrol; vegetable preservation

Citation: Zhu, R.; Liu, X.; Li, X.; Zeng, K.; Yi, L. Transformation of Inferior Tomato into Preservative: Fermentation by Multi-Bacteriocin Producing *Lactobacillus paracasei* WX322. *Foods* **2021**, *10*, 1278. https://doi.org/10.3390/foods10061278

Academic Editor: Peng Jin

Received: 20 April 2021
Accepted: 25 May 2021
Published: 3 June 2021

1. Introduction

Vegetables, as an essential food in people's daily life, can provide dietary fiber, minerals, vitamins, and other nutrients for the human body [1]. The health benefits of vegetable consumption include promoting gastrointestinal peristalsis [2], preventing chronic diseases (such as high blood pressure, diabetes, dementia) [3], boosting immunity [4], and so on. Experts of the United Nations Food and Agriculture Organization (FAO) and the World Health Organization (WHO) have encouraged people to eat more vegetables and fruits to reduce the risk of chronic diseases and cancer [5]. Global vegetable production is increasing, but it is still hard to meet the recommended intake (≥240 g/day according to WHO), especially in developing countries. For example, mean vegetable intake is 71 g/day in Melanesia [6], far less than the recommendation. According to the reports of FAO, a real challenge laid before the world's vegetable supply is postharvest loss and waste. Fruit and vegetable waste is one of the main food wastes, representing 0.5 billion tons per year (FAO 2011). In the US, 9% of fruit and 8% of vegetables are lost at the retail stage, and a further 19% of fruit and 22% of vegetables are not eaten at the consumption stage [7]. These wastes are not only critical at the agricultural stage in developing countries, but they are also high at the processing stage [8,9]. Waste from harvest inferior products and processing by-products accounts for about 30% of whole vegetables [10]. Therefore, innovative approaches and technology in combating postharvest loss and waste of vegetables are urgently needed.

Globally, postharvest loss of vegetables is as high as 60% (usually 20–60%) of total vegetable production [7]. Postharvest losses of vegetables are mainly caused by physiological disorders and infectious diseases. Infectious diseases of postharvest vegetables include bacterial diseases and fungal diseases, the former has caused more losses in leafy vegetables. Bacterial soft rot is a destructive disease, which causes more serious losses than any other bacterial disease [11]. For example, approximately 22% of potatoes are lost per year, among which bacterial soft rot alone accounts for 30–50% [12]. Bacterial pathogens leading to soft rot of vegetables include *Pectobacterium* [12], *Dickeya* [13], *Pseudomonas* [14], *Xanthomonas* [15], etc. *Pectobacterium carotovorum*, formerly *Erwinia carotovorum*, is the most commonly reported pathogen of vegetable soft rot and lists among the top 10 plant pathogenic bacteria [16]. Soft rot caused by *P. carotovorum* can occur anytime during postharvest handling, especially storage and transportation. *P. carotovorum* has a very broad range of hosts, including Solanaceae, Cruciferae, Cucurbitaceae, Compositae, Liliaceae, etc., involving almost all kinds of vegetables. The huge economic loss derived from soft rot has triggered great effort of researchers to find methods to control *P. carotovorum* in vegetables.

A lot of vegetable waste is generated during the processes of agricultural production, postharvest handling and storage, including inferior products and/or by-products, especially for the processing of fresh-cut vegetables [17]. Strategies of vegetable waste management have been reviewed by Plazzotta et al. [18], Esparza et al. [19], among others. Conventional waste management technologies consist of animal feeding, soil amendment, composting, and anaerobic digestion (biogas). Value-added applications contain extraction of bioactive compounds (e.g., flavonoids, phenolic acids, terpenes, oils, dietary fiber), production of enzymes, production of exopolysaccharides, production of biofuels, synthesis of bioplastics. Like the production of enzymes, vegetable wastes also can be used to produce antimicrobials through providing nutrition for the growth of producer microorganisms, which can be applied to control soft rot caused by *P. carotovorum*. However, few works have focused on such a transformation.

Bacteriocin is a kind of antibacterial polypeptide or precursor polypeptide synthesized by ribosomes in the process of bacteria metabolism [20]. It has ideal antibacterial properties and inhibits the growth of some food pathogenic and spoilage bacteria without side effects as a food preservative [21,22]. Nevertheless, bacteriocin is rarely reported to control soft rot of postharvest vegetables. In our previous study, a multi-bacteriocin producing lactic acid bacteria, *Lactobacillus paracasei* WX322, was isolated from fermented vegetable (pickle) [23] with good antibacterial activity against *P. carotovorum*. Bacteriocins from lactic acid bacteria are usually produced in Man-Rogosa-Sharpe broth with high cost. Pickle inspires us that vegetable wastes can be used to provide nutrition for the growth and bacteriocin production of *L. paracasei* WX322. Previous results indicated that fermentation products of *L. paracasei* WX322 in some vegetables could inhibit the growth of *P. carotovorum* (not published). The aims of this study were: to find appropriate vegetable waste to produce bacteriocins with higher activity than Man-Rogosa-Sharpe broth (to alternative to Man-Rogosa-Sharpe broth), then to analyze the growth property and bacteriocin production of *L. paracasei* WX322 in tomato juice, and finally to investigate the control potential of bacteriocin products from tomato juice fermentation to soft rot of different vegetables. This study can not only provide a new way to transform inferior tomato into value-added product (food preservative), but also provide a biocontrol approach with high safety to reduce postharvest losses of vegetables.

2. Materials and Methods

2.1. *Antibacterial Activity of L. paracasei WX322 Fermentation Products from 12 Kinds of Vegetables*

2.1.1. Screening of Vegetable Fermentation with Good Antibacterial Activity

L. paracasei WX322 (CGMCC No.17710) was preserved in our lab. *Pectobacterium cartovorum* subsp. *brasilliense* BZA12 (*Pcb* BZA12) was donated by Prof. Changyuan Liu from Liaoning Academy of Agricultural Sciences. Vegetables of tomato (*Solanum lycopersicum*), cucumber (*Cucumis sativus*), Chinese cabbage (*Brassica rapa* ssp. pekinensis), soybean sprouts (*Glycine max* L. Merrill), mungbean sprout (*Vigna radiate* L. Wilzcek), okra (*Abelmoschus esculentus* L. Moench), garlic chives (*Allium tuberosum* Rottl. ex Spreng), green bean (*Phaseolus vulgaris*), carrot (*Daucus carota*), balsam pear (*Momordica charantia*), white ground (*Benincasa hispida*), purple cabbage (*Brassica oleracea* L. var. Capitata) were collected from the local supermarket of Beibei, Chongqing, China.

In order to find a viable alternative to Man-Rogosa-Sharpe broth to produce bacteriocin, inferior products losing the commodity value of the 12 vegetables above were squeezed into juice. Specifically, vegetables were washed using running water, and then cut into small pieces using a knife. The pretreated vegetables were put into a Juicer WBL2501B (Midea, Guangdong, China) and squeezed with appropriate water. Juices of different vegetables were sterilized at 121 °C for 15 min. *L. paracasei* WX322 was inoculated into 100 mL sterile Man-Rogosa-Sharpe broth and cultured at 37 °C under 120 rpm for 24 h to logarithmic phase as seed. The seed of *L. paracasei* WX322 was inoculated into each prepared vegetable juice (100 mL) at a dose of 1%, and then stationarily cultured at 30 °C for 10 days. Subsequently, the fermentation product of each vegetable juice was centrifuged (5000 rpm, 15 min, 25 °C) using a high-speed freezing centrifuge ALLEGRA X-15R (Beckman Coulter Inc., CA, USA). The supernatant was filtered (0.22 μM pore size) and concentrated five folds in a vacuum freeze dryer LGJ-10 (Beijing Songyuan Huaxing Technology Develop Co., Ltd., Beijing, China). The antibacterial activity of the concentrated samples against *Pcb* BZA12 was determined by the agar well diffusion method as in our previous study [24]. The same treatment with Man-Rogosa-Sharpe broth was used as the control. All treatments were carried out for three times.

2.1.2. Verification of Antibacterial Activity from *L. paracasei* WX322 Metabolites

Components of some fruits and vegetables have been reported to possess antibacterial activity [25,26]. To eliminate the possibility that the antibacterial activity was from vegetable components, vegetable juice fermentation with and without *L. paracasei* WX322 were simultaneously undertaken. Tomato, Chinese cabbage, mungbean sprout, purple cabbage and okra were tested, and the fermentation process was carried out as described above. Supernatant without concentration was applied to test antibacterial activity, in which *Pcb* BZA12 was used as the indicator.

2.2. *Microbial Growth of L. paracasei WX322 in Tomato Juice and Antibacterial Activity Analysis*

The seed of *L. paracasei* WX322 was inoculated into prepared tomato juice at a ratio of 1%, and then incubated at 30 °C in stationary status. Sample was taken out each day, and then applied to analysis of microbial growth and antibacterial activity. Three replications of the tomato juice fermentation were conducted.

For microbial growth measurement of *L. paracasei* WX322 in tomato juice, the juice was mixed well and a 1 mL sample was taken out. The sample was diluted 10-fold using sterile saline (0.9% NaCl) from 10^{-1} to 10^{-8}. Each dilution was spread on Man-Rogosa-Sharpe agar plate. The plates were incubated at 37 °C for 48 h, and then colony number was counted.

For antibacterial activity measurement of fermentation supernatant, the juice was mixed well and 1 mL sample was taken out. The supernatant was obtained by centrifugation and its antibacterial activity was determined by the agar well diffusion method, in which *Pcb* BZA12 was used as the indicator.

2.3. Sensitivity of Fermentation Product of L. paracasei WX322 in Tomato Juice to Heat and Proteinases

L. paracasei WX322 was inoculated in sterile tomato juice and incubated at 30 °C for 10 days. The supernatant was obtained by centrifugation and subjected to assays of sensitivity analysis to heat and proteinases referring to Qi et al. [11].

For sensitivity analysis of fermentation product to heat, the sample was heated at 60 °C, 80 °C, 100 °C, 120 °C for 10 min, 20 min, 30 min, respectively. Among temperatures, 60 °C and 80 °C were conducted in a Digital Thermostat Water Bath HH-6 (LiChen, Shanghai Lichen Bangxi Instrument Technology Co., Ltd., Shanghai, China), 100 °C and 120 °C in a vertical autoclave G154D (Zealway, Xiamen Zealway Instrument Inc., Xiamen, China). After treatment, the samples were cooled to room temperature (about 20 °C), and the residual antibacterial activity was measured using the agar well diffusion method. The indicator used was *Pcb* BZA12. Unheated sample was used as the control.

For sensitivity analysis of the fermentation product to proteases, pepsin and proteinase K were used. Pepsin and proteinase K were dissolved in corresponding optimal buffers at 2 mg/mL: pH 3.0 citric acid-sodium citrate buffer, pH 7.0 Tris-HCl buffer, respectively. Then, the enzyme solution was added into samples at a final concentration of 1 mg/mL. After incubated at 37 °C (pepsin) or 55 °C (proteinase K) for 4 h, the residual antibacterial activity was determined using the agar well diffusion method. The indicator used was *Pcb* BZA12. Fermentation product without enzyme treatment and enzyme buffers without fermentation product were used as controls.

2.4. Preparation of Bacteriocin Sample and Its Antibacterial Activity Determination

L. paracasei WX322 was cultured in 3 L tomato juice at 30 °C for 10 days to prepare the bacteriocin sample. Subsequently, soluble polysaccharide from tomato was removed by alcohol precipitation [27,28]. Specifically, supernatant was obtained by centrifugation and ethanol was gradually added into the supernatant up to 80%. After stirred overnight, the precipitated saccharides were removed by centrifugation and ethanol in the supernatant was removed by rotary evaporation on a Rotary Evaporator RE-52AA (YaRong, Shanghai Yarong Biochemical instrument Factory, Shanghai, China) at 45 °C and 0.07 MPa. Moreover, microporous adsorption resin D101 was used to get rid of pigment in the sample. Antibacterial activity of the bacteriocin sample was measured using the agar well diffusion method after serial 2-fold dilution and expressed in arbitrary units (AU) per milliliter as described by Sabo et al. [29].

2.5. Scanning Electron Microscope (SEM) Observation

To investigate the effect of bacteriocin sample on cell morphology of *Pcb* BZA12, SEM was used referring to Li et al. [30]. Cells of *Pcb* BZA12 were collected in mid-logarithmic phase ($OD_{600nm} \approx 0.2$) by centrifugation at 5000 rpm for 15 min. Then, cells were resuspended in fresh LB broth and bacteriocin sample was added with a final concentration of 320 AU/mL. After treatment for 2 h at 37 °C, cells were washed with sterile saline three times to remove the residual bacteriocin sample. Subsequently, cells were fixed on glass slides using 2.5% (v/v) glutaraldehyde at 4 °C for 12 h. Further, cells were washed with sterile water and gradually dehydrated in ethanol solutions of 50%, 70%, 80%, 90%, 100%, and then in 100% acetone. Finally, the sample on the glass slide was dried completely and the morphology change of cells after bacteriocin sample treatment was observed by a Phenom Pro10102 scanning electron microscopy (Phenom-World, Eindhoven, The Netherlands). Areas with homogeneous cells were selected to take pictures at 10,000× g magnification. The same treatment with sterile saline instead of bacteriocin sample was used as the control.

2.6. Controlling Bacterial Soft Rot of Five Vegetables

Vegetables of Chinese cabbage, green bean, tomatoes, cucumber and balsam pear were purchased from local supermarket. Vegetables were treated as follows: Chinese cabbage and green bean were drilled on the surface with a sterile punch, tomatoes surface was peeled off in a small area, and cucumber and balsam pear were chopped to 5 cm in length. Bacterial cells of *Pcb* BZA12 were collected in mid-logarithmic phase and resuspended in sterile saline to OD_{600nm} = 0.2 using an Ultraviolet-visible Spectrophotometer V-5000 (METASH, Shanghai Yuanxi Instrument Co., Ltd., Shanghai, China). Then, 20 μL cell suspension of *Pcb* BZA12 was added into each well or onto the cut area. Vegetables were allowed to air dry at room temperature for 3 h. The vegetables inoculated with *Pcb* BZA12 were divided into two groups. For one group, 20 μL of 320 AU/mL bacteriocin sample was added into each well or onto the cut area. For the other group, 20 μL sterile water instead of bacteriocin sample was added as the control. Each group of each vegetable had 10 produces, which was conducted with 3 replications. All treated vegetables were allowed to air dry at room temperature for 3 h. Finally, vegetables were packaged in bags and placed at 30 °C to observe the incidence of soft rot.

2.7. Statistical Analysis

Three biological replicates were performed for all experiments and the results were expressed as mean ± standard deviation. Figures were carried out with OriginPro 8.1 (OriginLab, Northampton, MA, USA). Experimental differences were determined using a one-way analysis of variance (ANOVA) and Duncan test at 0.05 levels.

3. Results and Discussion

3.1. Screening Vegetable Alternative to Man-Rogosa-Sharpe Broth among 12 Kinds of Vegetables

P. carotovorum is generally known as a tissue maceration agent and causes bacterial soft rot disease [31], which has led to huge economic losses of postharvest vegetables. With the improvement of awareness of environmental protection and health, biological methods are more preferred by consumers to control soft rot of postharvest vegetables. Lactic acid bacteria are generally recognized as safe (GRAS) and have been used to combat with diseases of fruits and vegetables in both preharvest and postharvest in recent years. For example, *Lactobacillus plantarum* strain BY was used to control the soft rot of Chinese cabbage under field conditions [32] and *Lactobacillus sucicola* JT03 was used to control green mold of postharvest citrus [33]. Our previous study indicated that *L. paracasei* WX322 had good inhibitory activity against *P. carotovorum* in vitro and the bacteriocin from *L. paracasei* WX322 played a key role [11]. However, Man-Rogosa-Sharpe broth is not appropriate to be used directly in vegetables because it will affect the flavor and color of vegetables. Moreover, the cost of Man-Rogosa-Sharpe broth is high. Therefore, we aimed to seek low-cost and healthy nutrition provider for the production of bacteriocin by *L. paracasei* WX322.

Antibacterial activity of fermented products by *L. paracasei* WX322 from different vegetable juices after 5-fold concentration is shown in Table 1. Man-Rogosa-Sharpe broth was used as the reference whose inhibition zone diameter to *P. carotovorum* was 40.5 ± 0.4 mm. It is interesting that the fermentation products of two kinds of vegetables, namely tomato and cucumber, had higher inhibitory activity than Man-Rogosa-Sharpe broth, especially tomato with an inhibition zone diameter 47.3 ± 1.6 mm. The other 10 kinds of vegetables also had good antibacterial activity with inhibition zone diameter ranging from 37.4 ± 1.2 mm to 40.3 ± 0.4 mm. Lactic acid bacteria are usually a small part of the autochthonous microbiota of vegetables and fruits [34]. *L. paracasei* WX322 was isolated from homemade pickle [23], which is also an autochthonous lactic acid bacteria of vegetables. As we expected, vegetables can be used to provide nutrition for the bacteriocin production of *L. paracasei* WX322. Among 12 kinds of vegetables tested, tomato was the best nutrition provider. Tomato is grown and consumed all over the world with yield over 180 million tonnes per year according to the FAO. The generation of inferior products and by-products is inevitable during tomato planting and processing. However, more works are focused on value-added utilization of tomato pomace [35]. Attention also needs to be paid to the application of inferior tomato. In this study, production of bacteriocin by tomato juice fermentation from inferior products also provides another value-added utilization of tomato waste.

Table 1. Statistics of inhibition zone diameter of *L. paracasei* WX322 fermentation products from 12 kinds of vegetables after 5-fold concentration.

No.	Vegetables	Diameter of Inhibition Zone (mm) *
1	Tomato	47.3 ± 1.6^{a}
2	Cucumber	42.7 ± 1.2^{b}
3	Man-Rogosa-Sharpe broth	40.5 ± 0.4^{c}
4	Cabbage	40.3 ± 0.4^{cd}
5	Soybean sprout	39.9 ± 0.8^{cde}
6	Mungbean sprout	39.4 ± 1.8^{cde}
7	Okra	39.2 ± 0.6^{cdef}
8	Garlic chives	38.7 ± 1.0^{cdef}
9	Green bean	38.5 ± 1.3^{def}
10	Carrot	38.4 ± 0.8^{def}
11	Balsam pear	38.2 ± 0.9^{ef}
12	White ground	38.2 ± 0.5^{ef}
13	Purple cabbage	37.4 ± 1.2^{e}

* Inhibition zone diameter followed by different letters are significantly different by analysis of variance (ANOVA) and Tukey's test ($p < 0.05$).

Components of some fruits and vegetables also have antimicrobial activity [25,26]. Although great antibacterial activity of fermented products was observed in Table 1, it is unclear that whether the inhibition action was from metabolites produced by *L. paracasei* WX322 or from both components of vegetables and lactic acid bacteria metabolites. Therefore, antibacterial activity of five kinds of vegetable fermentations without *L. paracasei* WX322 was analyzed. As shown in Figure 1, the reference (A2, Man-Rogosa-Sharpe broth) and vegetables (B2, C2, D2, E2, F2) without *L. paracasei* WX322 had no inhibitory activity against *P. carotovorum*. In contrast, all fermentations with *L. paracasei* WX322 (A1, B1, C1, D1, E1, F1 in Figure 1) exhibited obvious inhibition zone. The result indicated that antibacterial activity of fermentation supernatant came from metabolites of *L. paracasei* WX322 not components of vegetables. Namely, vegetable juice only provided nutrition for the growth and bacteriocin production of *L. paracasei* WX322. Production of active peptides by tomato fermentation using microorganism has also been reported by other studies. For example, tomato waste was reported to produce antioxidant and ACE-inhibitory peptides by fermentation using *Bacillus subtilis* [36]. Therefore, the juice from inferior tomato (Figure 1G) was selected as medium for bacteriocin production by *L. paracasei* WX322 to alternative to Man-Rogosa-Sharpe broth.

3.2. Microbial Growth of L. paracasei WX322 and Antibacterial Activity of Fermented Product in Tomato Juice

In order to understand fermentation characteristics of *L. paracasei* WX322 in tomato juice, a dynamic process of microbial growth was investigated and the antibacterial activity of the fermentation product was measured. As shown in Figure 2A, the number of *L. paracasei* WX322 increased in the first 5 days and reached a maximum of about 10.19 $\log_{10}$ colony-forming units (cfu)/mL on the fifth day. Subsequently, a slight decrease of microbial number was observed, which may derive from the gradual consumption of nutrients and accumulation of metabolites. At the same time, tomato juice on the bottom was collected and its residues after centrifugation was observed. According to Figure 2C, residues of fermented tomato juice with *L. paracasei* WX322 was covered by white cells (left tube) while that without *L. paracasei* WX322 had no white (right tube). This result also revealed the great growth of *L. paracasei* WX322 in tomato juice.

Fermentation product of *L. paracasei* WX322 using tomato juice exhibited antibacterial activity on the second day (Figure 2B), which was close to our previous study using Man-

Rogosa-Sharpe broth in which bacteriocin was produced after fermentation for 36 h [11]. Moreover, with the extension of fermentation time, antibacterial activity of fermentation supernatant gradually increased. On the tenth day, the inhibition zone diameter of fermented tomato juice supernatant reached the maximum. Bacteriocins are usually secondary metabolites [37] and accumulate extracellularly with the cell density increase because of quorum sensing [38,39]. During the fermentation period of 10 days, the fermented product on the last day had the highest activity, therefore, 10-day was used as the fermentation time for tomato juice to produce bacteriocin by *L. paracasei* WX322.

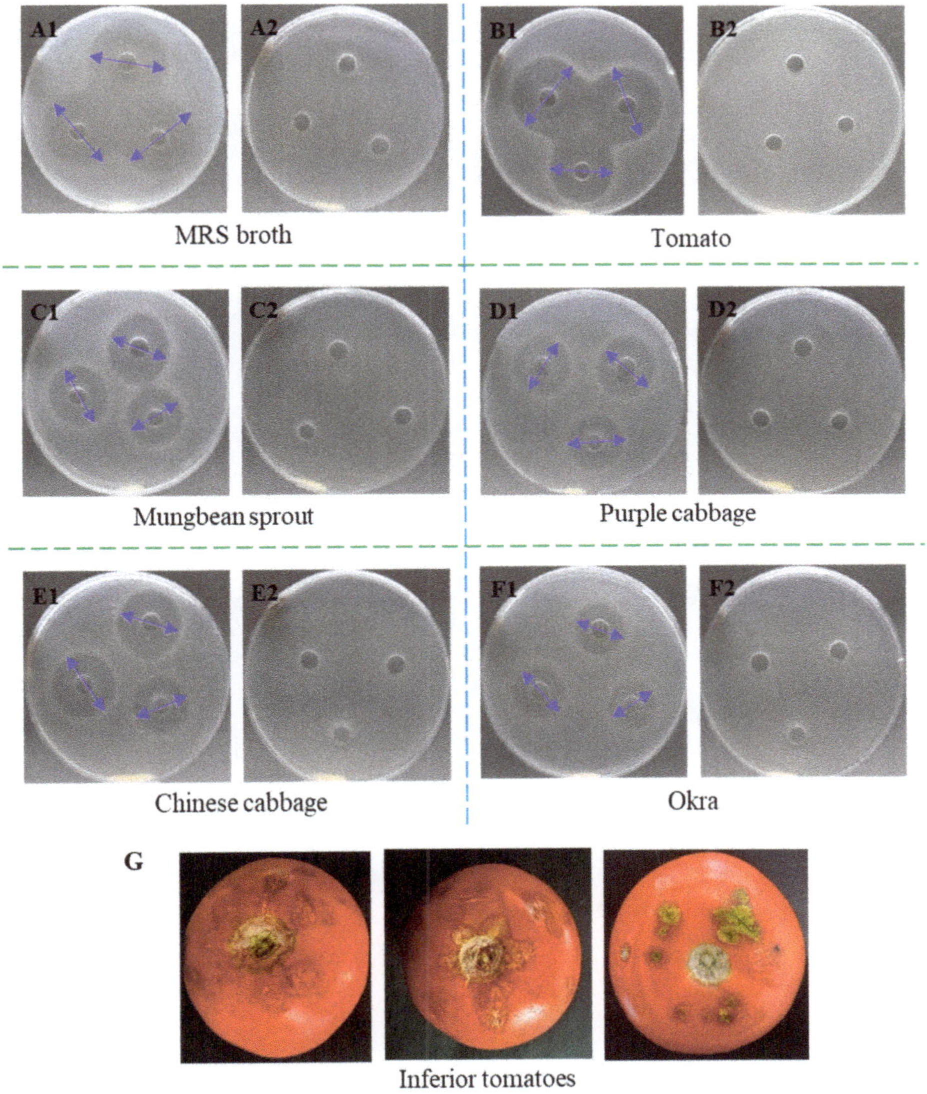

Figure 1. Antibacterial activity of fermentation supernatant. (**A1**), Man-Rogosa-Sharpe broth with *L. paracasei* WX322 as a reference; (**A2**), Man-Rogosa-Sharpe broth without *L. paracasei* WX322 as a reference. (**B1**), Tomato with *L. paracasei* WX322; (**B2**), tomato without *L. paracasei* WX322. (**C1**), Mungbean sprout with *L. paracasei* WX322; (**C2**), mungbean sprout without *L. paracasei* WX322. (**D1**), Purple cabbage with *L. paracasei* WX322; (**D2**), purple cabbage without *L. paracasei* WX322. (**E1**), Chinese cabbage with *L. paracasei* WX322; (**E2**), Chinese cabbage without *L. paracasei* WX322. (**F1**), Okra with *L. paracasei* WX322; (**F2**), okra without *L. paracasei* WX322. (**G**), Inferior products of tomato as an example. Inhibition zone diameter was marked by blue arrows.

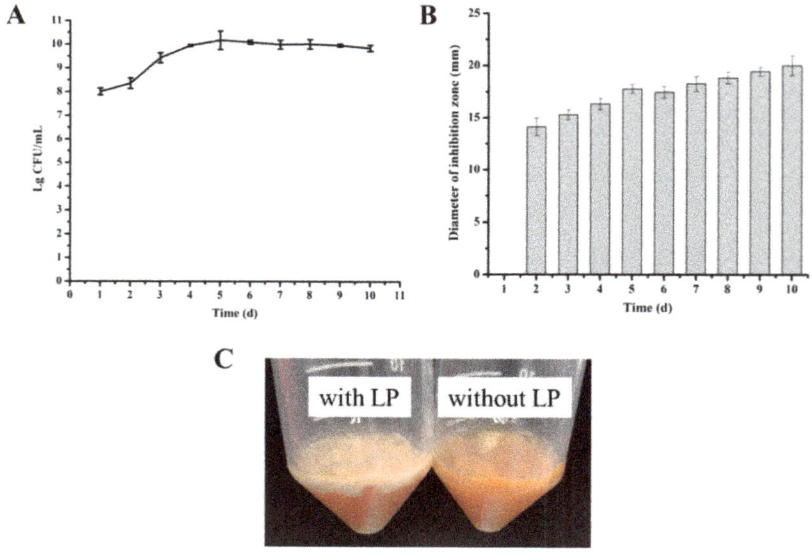

Figure 2. Fermentation characteristics of tomato juice by *L. paracasei* WX322. (**A**), Microbial number of *L. paracasei* WX322; (**B**), antibacterial activity of fermentation supernatant; (**C**), residues of fermented tomato juice after centrifugation, LP is *L. paracasei* WX322.

3.3. Sensitivity of Fermentation Product of L. paracasei WX322 from Tomato Juice to Heat and Proteinase

To make a further comparison of the bacteriocin produced by *L. paracasei* WX322 from Man-Rogosa-Sharpe broth and tomato juice, its sensitivity to heat and proteinases was analyzed. As shown in Figure 3A, the fermentation product from tomato juice had excellent thermostability that no significant difference was found in different time and temperature treatments. It still had good antibacterial activity even treated at 120 °C for 30 min. The excellent thermostability of bacteriocin of *L. paracasei* WX322 from tomato juice in this study remained the same as it was from Man-Rogosa-Sharpe broth in our previous study [11].

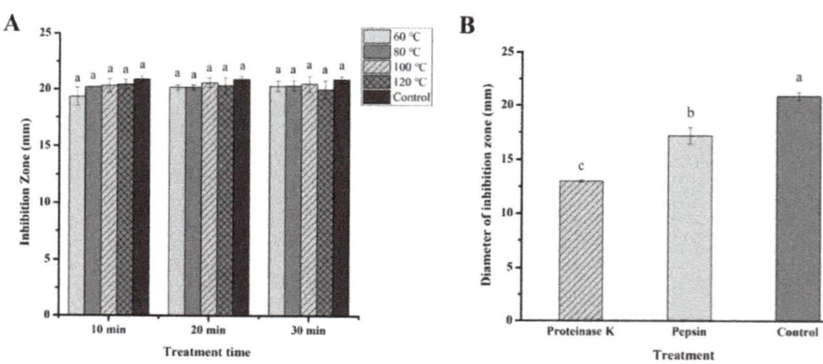

Figure 3. Sensitivity of fermentation product of *L. paracasei* WX322 from tomato juice to heat (**A**) and proteinase (**B**). (**A**), Fermentation product was heated at 60 °C, 80 °C, 100 °C, 120 °C for 10 min, 20 min, 30 min, respectively. (**B**), Fermentation product was treated by proteinases. Letters of a, b, c are significantly different by analysis of variance (ANOVA) and Tukey's test ($p < 0.05$).

At the same time, sensitivity of fermentation product from tomato juice to proteinase K and pepsin in this study was also in line with that from Man-Rogosa-Sharpe broth in our previous study [11]. Namely, treatments by proteinase K and pepsin caused partial loss of antibacterial activity.

In short, characteristics of the bacteriocin produced by *L. paracasei* WX322 from tomato juice are identical to those from Man-Rogosa-Sharpe broth. Therefore, tomato juice can be used to replace Man-Rogosa-Sharpe broth to produce bacteriocin by *L. paracasei* WX322.

3.4. Effect of Bacteriocin of L. paracasei WX322 from Tomato Juice on Cell Morphology of Pcb BZA12

A bacteriocin sample produced by *L. paracasei* WX322 was prepared by fermentation in tomato juice. According to the report from Williams and Bevenue [28], tomato solids insoluble in 80% ethyl alcohol are primarily polysaccharides. Therefore, 80% ethanol was used to remove polysaccharides from tomato. At the same time, some undesired pigments were removed by resin. Finally, antibacterial activity of the obtained bacteriocin sample was 640 AU/mL.

To observe the effect of the bacteriocin produced by *L. paracasei* WX322 from tomato juice on cell morphology of *Pcb* BZA12, 320 AU/mL sample was used to treat *Pcb* BZA12. As shown in Figure 4A, *Pcb* BZA12 cells of control had plump and complete outline. However, great deformation was observed after bacteriocin treatment (Figure 4B). Most cells exhibited a collapsed surface with a great wrinkle. Moreover, only a wizened envelope was observed for some cells. The great change of cell morphology indicates that the envelope integrity of *Pcb* BZA12 was damaged after the treatment of *L. paracasei* WX322 bacteriocin, followed by leakage of intracellular material. Damage of the cell envelope is an usual action mechanism of bacteriocins, such as Mutacin 1140 [40], AS-48 [41], and lacticin Q [42]. These bacteriocins can form pores on the cell membrane and lead to leakage of intracellular material. We infer that action mechanisms of the bacteriocin produced by *L. paracasei* WX322 to *Pcb* BZA12 is also involved in pore formation.

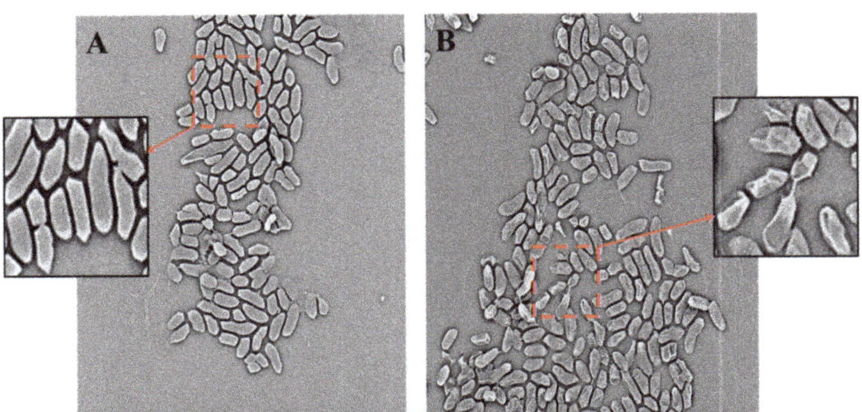

Figure 4. Scanning electron microscope (SEM) images of *Pcb* BZA12. (**A**), control with sterile saline; (**B**), treatment with bacteriocin of *L. paracasei* WX322 from tomato juice.

3.5. Controlling Bacterial Soft Rot of Five Kinds of Vegetables by Bacteriocin from Tomato Juice Fermentation

According to results above, bacteriocin of *L. paracasei* WX322 from tomato juice fermentation had good antibacterial activity against *Pcb* BZA12 in vitro. To further investigate its application potential as vegetable preservation in vivo, controlling effectiveness to soft rot of five vegetables were studied as shown in Figure 5. It can be found from Figure 5A,a that the bacteriocin treatment significantly inhibited the soft rot of Chinese cabbage. For the control, incidence of soft rot was 100% on the first day (Figure 5A1), then area of soft rot increased on the second day (Figure 5A2), followed by rotten leaves losing integrity and generating foul smell on the third day (Figure 5A3). However, for bacteriocin treatment, most leaves kept fresh and intact on the first day (Figure 5a1) and second day (Figure 5a2) with disease incidence of 10% ± 0% and 16.66% ± 4.71%, respectively. Incidence of soft rot was only 20% ± 8.16% on the third day (Figure 5a3). Bacterial soft rot disease, caused by *P. cartovorum*, is a fatal disease of postharvest cabbage throughout the world. Some antagonistic bacteria have been reported to control soft rot of cabbage [43], including bacteriocin-producing bacteria [44]. But the control potential of bacteriocin to soft rot of postharvest cabbage is unknown. This study can provide the evidence of biocontrol of cabbage soft rot using bacteriocin.

P. cartovorum is also an important pathogen causing soft rot of cucumber [45]. The soft rot pathogen, *Pcb* BZA12, used in this study was isolated from diseased cucumber [46]. Therefore, controlling soft rot of cucumber by bacteriocin was also studied. As shown in Figure 5B,b, bacteriocin from tomato juice also had a good controlling effect on soft rot of cucumber. For the control, rot and yellowing of cucumber sections occurred 100% on the first day (Figure 5B1), and symptoms aggravated on the second day (Figure 5B2) and third day (Figure 5B3). In contrast, cucumber sections kept green without rot symptom all the time during the period of 3 days (Figure 5b1–b3) after being treated by the bacteriocin sample. The result indicates that bacteriocin of *L. paracasei* WX322 from tomato juice also can be used to control soft rot of cucumber.

Soft rot disease caused by *P. carotovorum* has led to significant postharvest losses to tomatoes [47]. *P. carotovorum* is not able to penetrate the surface directly, and therefore the tomato surface was peeled off in a small area to create a wound and then the pathogen was inoculated. For the control (Figure 5C), rotten rate of tomato was 3.33% ± 4.71%, 40% ± 0%, and 70% ± 14.14% on the first, second, and third day, respectively. During which the deterioration of soft rot, symptoms of water soak and sunken tissue were increasingly serious. The peel of some tomatoes even burst (Figure 5C3). However, for bacteriocin treatment (Figure 5c), no rotten tomato was found on the first two days and disease incidence was only 13.33% ± 9.43% on the third day. The result indicates that the bacteriocin of *L. paracasei* WX322 from tomato juice performs well on controlling soft rot of postharvest tomato.

P. carotovorum has been reported to cause soft rot of balsam pear [48]. In this study, the result of the control (Figure 5D) demonstrated that *P. carotovorum* can cause rapid and dissolving damage to balsam pear that this vegetable rotted into pus with a terrible foul smell on the third day (Figure 5D3). Bacteriocin treatment delayed the rot to a certain degree (Figure 5d) with alleviated symptoms.

As shown in Figure 5E, *P. carotovorum* caused rot and browning to green beans with a rotten rate of 76.67% ± 4.71% for the control on the third day. On the fourth day, disease incidence reached 100% with fragmentary beans (Figure 5E4). In contrast, all green beans had intact profile and rotten rate was 26.67% ± 4.71% and 40% ± 0% on the third day (Figure 5e3) and fourth day (Figure 5e4), respectively. The result indicates that bacteriocin of *L. paracasei* WX322 from tomato juice can significantly inhibit the soft rot of green bean caused by *P. carotovorum*.

Controls of the five vegetables (Figure 5A–E) demonstrated that *P. carotovorum* can multiply in the plant tissues where it causes water soak, sinking, softening, yellowing/browning, foul smell, etc. resulting in losses of commodity value and edible value. Currently, there are few effective ways to decontaminate infected vegetables that do not pose concerns to human health [48]. However, in this study, the bacteriocin of *L. paracasei* WX322 can significantly delay the soft rot of vegetables, especially for cabbage and cucumber. Moreover, the bacteriocin was fermented in tomato juice, which is safe and healthy.

Figure 5. Controlling soft rot of vegetables. (**A**,**a**), Chinese cabbage; (**A1–A3**) are controls on 1, 2, and 3 day, respectively; (**a1–a3**) are treatments with bacteriocin of *L. paracasei* WX322 on 1, 2, and 3 day, respectively. (**B**,**b**), cucumber; (**B1–B3**) are controls on 1, 2, and 3 day, respectively; (**b1–b3**) are treatments with bacteriocin of *L. paracasei* WX322 on 1, 2, and 3 day, respectively. (**C**) and (**c**), tomato; C1, C2, and C3 are controls on 1, 2, and 3 day, respectively; (**c1–c3**) are treatments with bacteriocin of *L. paracasei* WX322 on 1, 2, and 3 day, respectively. (**D**,**d**), balsam pear; (**D1–D3**) are controls on 1, 2, and 3 day, respectively; (**d1–d3**) are treatments with bacteriocin of *L. paracasei* WX322 on 1, 2, and 3 day, respectively. (**E**,**e**), green bean; (**E1–E4**) are controls on 1, 2, 3, and 4 day, respectively; (**e1–e4**) are treatments with bacteriocin of *L. paracasei* WX322 on 1, 2, 3, and 4 day, respectively.

4. Conclusions

P. cartovorum is one of the most important soft rot pathogens causing great postharvest losses worldwide. Controlling soft rot is vital for the extension of vegetable shelf-life. In this study, all the 12 kinds of vegetables tested can provide nutrition for the *L. paracasei* WX322 to produce antibacterial bacteriocin, among which tomato performed best. In tomato juice, *L. paracasei* WX322 grew well with the maximum colony number around 10.19 $\log_{10}$ cfu/mL on the fifth day and the highest production of bacteriocin on the tenth day. Tomato juice can be used to replace Man-Rogosa-Sharpe broth to produce bacteriocin by *L. paracasei* WX322 because characteristics of the bacteriocin from tomato juice and from Man-Rogosa-Sharpe broth were the same. The bacteriocin killed *P. carotovorum* by causing the collapse and significant deformation of cells according to SEM images. In vivo assay, the bacteriocin of *L. paracasei* WX322 fermented in tomato juice inhibited incidence of soft rot of vegetables in different degree. This study suggests that producing bacteriocin using tomato waste (by-products and/or inferior products) can not only provide an innovative method to utilize vegetable waste, but also provide a biocontrol method to control soft rot of some postharvest vegetables.

Author Contributions: Conceptualization, L.Y. and K.Z.; methodology, R.Z.; software, R.Z. and X.L. (Xiaoqing Liu); validation, L.Y. and K.Z.; formal analysis, X.L. (Xiaoqing Liu); investigation, X.L. (Xiaofen Li); resources, L.Y.; data curation, L.Y.; writing—original draft preparation, R.Z.; writing—review and editing, L.Y.; visualization, R.Z.; supervision, L.Y. and K.Z.; project administration, L.Y.; funding acquisition, L.Y. All authors have read and agreed to the published version of the manuscript.

Funding: This research was supported by the Natural Science Foundation of Chongqing (cstc2019jcyj-msxmX0357).

Acknowledgments: This research was supported by the Natural Science Foundation of Chongqing (cstc2019jcyj-msxmX0357). The authors thank Changyuan Liu from Liaoning Academy of Agricultural Sciences for his kind donation of *Pectobacterium cartovorum* subsp. *brasilliense* BZA12 (*Pcb* BZA12).

Conflicts of Interest: The authors declare no conflict of interest.

References

1. Gebbers, J.-O. Atherosclerosis, cholesterol, nutrition, and statins—A critical review. *GMS Ger. Med. Sci.* **2007**, *5*, Doc04. [PubMed]
2. Jin, W.; Chen, X.; Huo, Q.; Cui, Y.; Yu, Z.; Yu, L. Aerial parts of maca (*Lepidium meyenii* Walp.) as functional vegetables with gastrointestinal prokinetic efficacy in vivo. *Food Funct.* **2018**, *9*, 3456–3465. [CrossRef] [PubMed]
3. Jiang, X.; Huang, J.; Song, D.; Deng, R.; Wei, J.; Zhang, Z. Increased consumption of fruit and vegetables is related to a reduced risk of cognitive impairment and dementia: Meta-analysis. *Front. Aging Neurosci.* **2017**, *9*, 18. [CrossRef] [PubMed]
4. Park, K.-Y.; Jeong, J.-K.; Lee, Y.-E.; Daily, J.W. Health benefits of kimchi (Korean fermented vegetables) as a probiotic food. *J. Med. Food* **2014**, *17*, 6–20. [CrossRef] [PubMed]
5. Ramos, B.; Miller, F.; Brandão, T.; Teixeira, P.; Silva, C. Fresh fruits and vegetables—An overview on applied methodologies to improve its quality and safety. *Innov. Food Sci. Emerg. Technol.* **2013**, *20*, 1–15. [CrossRef]
6. Kalmpourtzidou, A.; Eilander, A.; Talsma, E.F. Global vegetable intake and supply compared to recommendations: A systematic review. *Nutrients* **2020**, *12*, 1558. [CrossRef] [PubMed]
7. Porat, R.; Lichter, A.; Terry, L.; Harker, R.; Buzby, J. Postharvest losses of fruit and vegetables during retail and in consumers' homes: Quantifications, causes, and means of prevention. *Postharvest Biol. Technol.* **2018**, *139*, 135–149. [CrossRef]
8. Salim, N.S.M.; Singh, A.; Raghavan, V. Potential utilization of fruit and vegetable wastes for food through drying or extraction techniques. *Nov. Tech. Nutr. Food Sci.* **2017**, *1*, 1–12. [CrossRef]
9. Ferreira, M.S.L.; Santos, M.C.P.; Moro, T.M.A.; Basto, G.J.; Andrade, R.M.S.; Goncalves, E. Formulation and characterization of functional foods based on fruit and vegetable residue flour. *J. Food Sci. Technol.* **2015**, *52*, 822–830. [CrossRef]
10. Sagar, N.A.; Pareek, S.; Sharma, S.; Yahia, E.M.; Lobo, M.G. Fruit and vegetable waste: Bioactive compounds, their extraction, and possible utilization. *Compr. Rev. Food Sci. Food Saf.* **2018**, *17*, 512–531. [CrossRef]
11. Qi, T.; Wang, S.; Deng, L.; Yi, L.; Zeng, K. Controlling pepper soft rot by *Lactobacillus paracasei* WX322 and identification of multiple bacteriocins by complete genome sequencing. *Food Control* **2021**, *121*, 107629. [CrossRef]
12. Viswanath, H.; Rather, R.A.; Bhat, K.; Peerzada, S. Management of post-harvest bacterial soft rot of potato caused by *Pectobacterium carotovorum* using bacterial bio-agents. *Pharma Innov. J.* **2020**, *9*, 31–35.

13. Hadizadeh, I.; Peivastegan, B.; Hannukkala, A.; Van Der Wolf, J.M.; Nissinen, R.; Pirhonen, M. Biological control of potato soft rot caused by *Dickeya solani* and the survival of bacterial antagonists under cold storage conditions. *Plant Pathol.* **2018**, *68*, 297–311. [CrossRef]
14. Godfrey, S.A.C.; Marshall, J.W. Identification of cold-tolerant *Pseudomonas viridiflava* and *P. marginalis* causing severe carrot postharvest bacterial soft rot during refrigerated export from New Zealand. *Plant Pathol.* **2002**, *51*, 155–162. [CrossRef]
15. Lo, H.-H.; Liao, C.-T.; Li, C.-E.; Chiang, Y.-C.; Hsiao, Y.-M. The clpX gene plays an important role in bacterial attachment, stress tolerance, and virulence in *Xanthomonas campestris* pv. *campestris*. *Arch. Microbiol.* **2019**, *202*, 597–607. [CrossRef]
16. Mansfield, J.; Genin, S.; Magori, S.; Citovsky, V.; Sriariyanun, M.; Ronald, P.; Dow, M.; Verdier, V.; Beer, S.V.; Machado, M.A.; et al. Top 10 plant pathogenic bacteria in molecular plant pathology. *Mol. Plant Pathol.* **2012**, *13*, 614–629. [CrossRef]
17. Wiryawan, F.S.; Marimin; Djatna, T. Value chain and sustainability analysis of fresh-cut vegetable: A case study at SSS Co. *J. Clean. Prod.* **2020**, *260*, 121039. [CrossRef]
18. Plazzotta, S.; Manzocco, L.; Nicoli, M.C. Fruit and vegetable waste management and the challenge of fresh-cut salad. *Trends Food Sci. Technol.* **2017**, *63*, 51–59. [CrossRef]
19. Esparza, I.; Jiménez-Moreno, N.; Bimbela, F.; Azpilicueta, C.A.; Gandía, L.M. Fruit and vegetable waste management: Conventional and emerging approaches. *J. Environ. Manag.* **2020**, *265*, 110510. [CrossRef]
20. Yi, L.; Dang, Y.; Wu, J.; Zhang, L.; Liu, X.; Liu, B.; Zhou, Y.; Lu, X. Purification and characterization of a novel bacteriocin produced by *Lactobacillus crustorum* MN047 isolated from koumiss from Xinjiang, China. *J. Dairy Sci.* **2016**, *99*, 7002–7015. [CrossRef]
21. Garcia-Gutierrez, E.; O'Connor, P.M.; Colquhoun, I.J.; Vior, N.M.; Rodríguez, J.M.; Mayer, M.J.; Cotter, P.D.; Narbad, A. Production of multiple bacteriocins, including the novel bacteriocin gassericin M, by *Lactobacillus gasseri* LM19, a strain isolated from human milk. *Appl. Microbiol. Biotechnol.* **2020**, *104*, 3869–3884. [CrossRef] [PubMed]
22. Lv, X.; Miao, L.; Ma, H.; Bai, F.; Lin, Y.; Sun, M.; Li, J. Purification, characterization and action mechanism of plantaricin JY22, a novel bacteriocin against *Bacillus cereus* produced by *Lactobacillus plantarum* JY22 from golden carp intestine. *Food Sci. Biotechnol.* **2018**, *27*, 695–703. [CrossRef] [PubMed]
23. Yi, L.; Qi, T.; Hong, Y.; Deng, L.; Zeng, K. Screening of bacteriocin-producing lactic acid bacteria in Chinese homemade pickle and dry-cured meat, and bacteriocin identification by genome sequencing. *LWT Food Sci. Technol.* **2020**, *125*, 109177. [CrossRef]
24. Yi, L.; Qi, T.; Ma, J.; Zeng, K. Genome and metabolites analysis reveal insights into control of foodborne pathogens in fresh-cut fruits by *Lactobacillus pentosus* MS031 isolated from Chinese Sichuan Paocai. *Postharvest Biol. Technol.* **2020**, *164*, 111150. [CrossRef]
25. Vallejo, C.V.; Minahk, C.J.; Rollán, G.C.; Rodríguez-Vaquero, M.J. Inactivation of *Listeria monocytogenes* and *Salmonella Typhimurium* in strawberry juice enriched with strawberry polyphenols. *J. Sci. Food Agric.* **2021**, *101*, 441–448. [CrossRef]
26. Masoud, K. Antibacterial properties of natural compounds extracted from plants compared to chemical preservatives against *Salmonella* spp.—A Systematic Review. *Int. J. Food Nutr. Saf.* **2017**, *8*, 13–31.
27. Ai, C.; Meng, H.; Lin, J.; Zhang, T.; Guo, X. Combined membrane filtration and alcohol-precipitation of alkaline soluble polysaccharides from sugar beet pulp: Comparision of compositional, macromolecular, and emulsifying properties. *Food Hydrocoll.* **2020**, *109*, 106049. [CrossRef]
28. Williams, K.T.; Bevenue, A. Vegetable components, some carbohydrate components of tomato. *J. Agric. Food Chem.* **1954**, *2*, 472–474. [CrossRef]
29. Sabo, S.; Converti, A.; Ichiwaki, S.; Oliveira, R.P. Bacteriocin production by *Lactobacillus plantarum* ST16Pa in supplemented whey powder formulations. *J. Dairy Sci.* **2019**, *102*, 87–99. [CrossRef]
30. Li, T.; Liu, Q.; Wang, D.; Li, J. Characterization and antimicrobial mechanism of CF-14, a new antimicrobial peptide from the epidermal mucus of catfish. *Fish Shellfish. Immunol.* **2019**, *92*, 881–888. [CrossRef]
31. Hajian-Maleki, H.; Baghaee-Ravari, S.; Moghaddam, M. Efficiency of essential oils against *Pectobacterium carotovorum* subsp. *carotovorum* causing potato soft rot and their possible application as coatings in storage. *Postharvest Biol. Technol.* **2019**, *156*, 110928. [CrossRef]
32. Tsuda, K.; Tsuji, G.; Higashiyama, M.; Ogiyama, H.; Umemura, K.; Mitomi, M.; Kubo, Y.; Kosaka, Y. Biological control of bacterial soft rot in Chinese cabbage by *Lactobacillus plantarum* strain BY under field conditions. *Biol. Control* **2016**, *100*, 63–69. [CrossRef]
33. Ma, J.; Hong, Y.; Deng, L.; Yi, L.; Zeng, K. Screening and characterization of lactic acid bacteria with antifungal activity against *Penicillium digitatum* on citrus. *Biol. Control* **2019**, *138*, 104044. [CrossRef]
34. Di Cagno, R.; Coda, R.; De Angelis, M.; Gobbetti, M. Exploitation of vegetables and fruits through lactic acid fermentation. *Food Microbiol.* **2013**, *33*, 1–10. [CrossRef]
35. Lu, Z.; Wang, J.; Gao, R.; Ye, F.; Zhao, G. Sustainable valorisation of tomato pomace: A comprehensive review. *Trends Food Sci. Technol.* **2019**, *86*, 172–187. [CrossRef]
36. Moayedi, A.; Mora, L.; Aristoy, M.C.; Safari, M.; Hashemi, M.; Toldrá, F. Peptidomic analysis of antioxidant and ACE-inhibitory peptides obtained from tomato waste proteins fermented using *Bacillus subtilis*. *Food Chem.* **2018**, *250*, 180–187. [CrossRef]
37. Kum, E.; Ince, E. Genome-guided investigation of secondary metabolites produced by a potential new strain *Streptomyces* BA2 isolated from an endemic plant rhizosphere in Turkey. *Arch. Microbiol.* **2021**, 1–8. [CrossRef]
38. Russell, J.B.; Mantovani, H.C. The bacteriocins of ruminal bacteria and their potential as an alternative to antibiotics. *J. Mol. Microbiol. Biotechnol.* **2002**, *4*, 347–355.
39. Hols, P.; Ledesma, L.; Gabant, P.; Mignolet, J. Mobilization of microbiota commensals and their bacteriocins for therapeutics. *Trends Microbiol.* **2019**, *27*, 690–702. [CrossRef]

40. Pokhrel, R.; Bhattarai, N.; Baral, P.; Gerstman, B.S.; Park, J.H.; Handfield, M.; Chapagain, P.P. Molecular mechanisms of pore formation and membrane disruption by the antimicrobial lantibiotic peptide Mutacin 1140. *Phys. Chem. Chem. Phys.* **2019**, *21*, 12530–12539. [CrossRef]
41. Perez, R.H.; Zendo, T.; Sonomoto, K. Circular and leaderless bacteriocins: Biosynthesis, mode of action, applications, and prospects. *Front. Microbiol.* **2018**, *9*, 2085. [CrossRef] [PubMed]
42. Vasilchenko, A.S.; Valyshev, A.V. Pore-forming bacteriocins: Structural–functional relationships. *Arch. Microbiol.* **2018**, *201*, 147–154. [CrossRef] [PubMed]
43. Cui, W.; He, P.; Munir, S.; He, P.; He, Y.; Li, X.; Yang, L.; Wang, B.; Wu, Y.; He, P. Biocontrol of soft rot of Chinese cabbage using an endophytic bacterial strain. *Front. Microbiol.* **2019**, *10*, 1471. [CrossRef]
44. Kyeremeh, A.G.; Kikumoto, T.; Chuang, D.-Y.; Gunji, Y.; Takahara, Y.; Ehara, Y. Biological control of soft rot of Chinese cabbage using single and mixed treatments of bacteriocin-producing avirulent mutants of *Erwinia carotovora* subsp. *carotovora*. *J. Gen. Plant Pathol.* **2000**, *66*, 264–268. [CrossRef]
45. Onkendi, E.M.; Ramesh, A.M.; Kwenda, S.; Naidoo, S.; Moleleki, L.N. Draft genome sequence of a virulent *Pectobacterium carotovorum* subsp. *brasiliense* isolate causing soft rot of cucumber. *Genome Announc.* **2016**, *4*, e01530-15. [CrossRef]
46. Huang, Y.; Liu, C.; Wang, H.; Guan, T.; Liu, L.; Yu, S. Bioinformatic analysis of the complete genome sequence of *Pectobacterium carotovorum* subsp. *brasiliense* BZA12 and candidate effector screening. *J. Plant Pathol.* **2018**, *101*, 39–49. [CrossRef]
47. Ahmed, F.A.; Arif, M.; Alvarez, A.M. Antibacterial effect of potassium tetraborate tetrahydrate against soft rot disease agent *Pectobacterium carotovorum* in tomato. *Front. Microbiol.* **2017**, *8*, 1728. [CrossRef]
48. Kubo, H.; Kanehashi, K.; Shinohara, H.; Negishi, H.; Matsuyama, N.; Suyama, K. Bacterial soft rot, a new disease of balsam pear (*Momordica charantia* L.) caused by *Erwinia carotovora* subsp. *carotovora*. *Jpn. J. Phytopathol.* **2009**, *75*, 173–175. [CrossRef]

foods

MDPI

Article

Inhibition of Three Citrus Pathogenic Fungi by Peptide PAF56 Involves Cell Membrane Damage

Wenjun Wang [1], Guirong Feng [1], Xindan Li [1], Changqing Ruan [1,2], Jian Ming [1,2] and Kaifang Zeng [1,2,3,*]

[1] College of Food Science, Southwest University, Chongqing 400715, China; wangwenjun_w@outlook.com (W.W.); fengguirong0128@outlook.com (G.F.); 18227589580@163.com (X.L.); changqing.r@hotmail.com (C.R.); mingjian1972@163.com (J.M.)

[2] Research Center of Food Storage & Logistics, Southwest University, Chongqing 400715, China

[3] Key Laboratory of Plant Hormones and Development Regulation of Chongqing, Chongqing 401331, China

* Correspondence: zengkaifang@hotmail.com

Abstract: The peptide PAF56 (GHRKKWFW) was reported to be an effective control for the main diseases of citrus fruit during postharvest storage. However, the mechanism of action of PAF56 is still unknown. In this paper, PAF56 might not induce defense resistance of citrus fruit. The SEM results visually indicated that the fungi mycelia became shrunken and distorted after being treated with PAF56. The destructive effects of PAF56 on the mycelial cell membrane of three kinds of pathogenic fungi (*Penicillium digitatum*, *Penicillium italicum*, and *Geotrichum citri-aurantii*) were verified by the K^+ leakage and the release of nucleic acid. Furthermore, the interaction between peptide PAF56 and the pathogen spores was investigated, including the changes in cell membrane permeability and dynamic observation of the interaction of fluorescein labeled TMR-PAF56 and *Geotrichum candidum* spores. The results indicated that the antifungal activity of PAF56 on spores was time-dependent and directly related to the membrane damage. This research provided useful references for further research and practical application of peptides.

Keywords: peptide PAF56; citrus fruit; changes cell structure; spores; membrane permeability

Citation: Wang, W.; Feng, G.; Li, X.; Ruan, C.; Ming, J.; Zeng, K. Inhibition of Three Citrus Pathogenic Fungi by Peptide PAF56 Involves Cell Membrane Damage. *Foods* **2021**, *10*, 2031. https://doi.org/10.3390/foods10092031

Academic Editor: Rotimi Aluko

Received: 23 July 2021
Accepted: 27 August 2021
Published: 29 August 2021

1. Introduction

Citrus is the type of fruit crop with the highest production worldwide, and the citrus industry has great economic importance. *Penicillium digitatum* (green mold), *P. italicum* (blue mold), and *Geotrichum citri-aurantii* (sour rot) are well known as the predominant citrus pathogens causing postharvest diseases during fruit storing and transportation. Recently, researchers have been trying to use various safe and effective approaches to control these diseases.

Antimicrobial peptides (AMPs) were widely studied as novel antibiotics and have been applied for controlling phytopathogens in agriculture, postharvest conservation, medical industry, and so on. Increasing antimicrobial peptides has been proved to be able to control infectious diseases of fruit and vegetables such as citrus and tomato [1–3]. In particular, short-chain cationic antimicrobial peptides attract the attention of researchers due to their cheap synthesis costs and excellent antimicrobial efficiency. Markedly, PAFs were a group of de novo designed hexapeptides with a good controlling effect against plant filamentous fungi [4]. PAF26 (RKKWFW) was reported to be an effective inhibitor for the growth of *P. digitatum* in vivo and in vitro [5,6], without lytic or cytotoxic effects on human cells. Further studies have shown that PAF26 has multiple effects on *P. digitatum* that ultimately result in permeation and killing [7,8]. When the N-terminal of PAF26 was extended by glycine and histidine residues (GH), PAF56 was obtained. It has been proved that PAF56 could control several pathogens, including fungi *P. digitatum*, *Fusarium oxysporum*, and Gram-negative bacterium *Escherichia coli* [9]. In our previous research, PAF56 exhibited an effective control on green and blue mold and sour rot in citrus fruit without a hemolytic effect. PAF56 could change the selective permeability of *P. digitatum*, *P. italicum*, and *G. candidum* mycelia after 48 h treatment, while SG (SYTOX Green) would enter the cell and bind to the nucleic acid, and emitted strong green fluorescence. The extracellular conductivity significantly increased with the increasing concentration of PAF56 [10]. The fungicidal mechanism of the control of PAF56 for the diseases of citrus fruit has not been revealed. It is not clear whether the mechanism of the control of active peptide PAF56 for those diseases is related to induced fruit defense resistance. In addition, fungi initiate their infection by disseminating spores, and then spores swell and germinate into hyphae, which results in severe yield loss in the citrus industry.

In the present study, we studied the effect of PAF56 on spores, and the mechanism related to cell membrane was further explored focusing on two aspects of fungal spores and mycelia.

2. Materials and Methods

2.1. Synthetic Peptide

Peptides were purchased from GenScript Corporation (Nanjing, China) at >90% purity. PAF56 (GHRKKWFW) was synthesized by the solid-phase method using 9-fluorenylmetho xycarbonyl (Fmoc)-type chemistry. TMR-PAF56 (PAF56 labeled with tetramethyl-rhodamine) modified covalently at its N-terminus was also synthesized. Stock solutions of PAF56 at 1 mmol L^{-1} were prepared in sterile ultrapure water, and stock solutions of TMR-PAF56 at 4 mmol L^{-1} were prepared in 5 mmol L^{-1} 3-(N-morpholino)-propane sulfonic acid (MOPS) and pH 7 buffer and stored in low-light conditions at −40 °C.

2.2. Fungal Strains

P. digitatum, *P. italicum*, and *G. candidum* were all isolated from the surface of naturally infected citrus fruit and identified by morphology and sequence of the internal transcribed spacer (ITS) rDNA region. The pathogens were purified and cultured in potato dextrose agar (PDA) plate at 25 °C [11]. Spores were obtained from 7-days-old plates and washed with sterile distilled water containing 0.1 g kg^{-1} Tween-80. Spores were titrated with a hemacytometer.

2.3. Fruit and Treatment

Citrus fruit [*Citrus sinensis* (L.) Osbeck] were harvested at their commercial maturity from a local orchard (Beibei, Chongqing). After harvesting, the fruit were selected based on uniform size, color, and absence of defects. The fruit were surface-disinfected with 2% (*v*/*v*) sodium hypochlorite for 2 min, washed with water, and air-dried at room temperature (20 °C).

To test whether defense resistance is induced in citrus by PAF56, two holes (3 mm × 4 mm) were drilled at two sites around the equator of each fruit. PAF56 (10 μL, 64 μmol L^{-1}) was pipetted into each wound site. Citrus fruit inoculated with ultrapure water only was set as the control. After 24 h, a 10 μL suspension of 1×10^4 CFU mL^{-1} of fungi (*P. italicum*, *P. digitatum*, or *G. candidum*) spores was inoculated into a new rind site that was 1 cm away from the initial treated point. Three replicates (15 fruits per replicate, 2 wounds per fruit) were prepared for each group. All fruits were stored at 25 °C and 90% RH (relative humidity). The disease incidence (DI) and the lesion diameter (LD) were assessed daily. The mean values ± S.D. of DI and LD were calculated [12].

2.4. Scanning Electron Microscope (SEM) Analysis

We observed the damage of PAF56 to the mycelia morphology of the fungi by scanning electron microscope (SEM). Mycelia cultured in potato dextrose broth (PDB) liquid for 2 d were collected, washed, and resuspended in the PAF56 solutions (0, 10, and 100 μmol L^{-1}) for 48 h. Mycelia morphology of the three fungi was observed according to the previous method with a minor modification [13], by using a JEOL JSM-6510LV SEM (JEOL, Tokyo, Japan) operating at 25 kV.

2.5. Measurement of Indicators Related to Change of Permeability of Cell Membrane

The efflux of K^+ of cytoplasmic components is an important indicator of the increasing permeability of the cell membrane. PAF56 solutions (10 or 100 μmol L^{-1}) used in each treatment group were prepared, while controls without PAF56 were treated similarly. The concentration of extracellular potassium in the supernatant and release of cytoplasmic constituents were measured by flame atomic absorption spectroscopy (Shimadzu AA6300, Kyoto, Japan) and using a Multiskan Spectrum microplate spectrophotometer at 260 nm (BioTek Instruments, Inc., Winooski, VT, USA), respectively [12,14,15]. The mycelia were collected after shaking at 25 °C for 2 d and washed before resuspension in sterilized distilled water (for the measurement of extracellular potassium concentration) or phosphate buffer (0.05 mol L^{-1}, phosphate, pH 7.0) (for the measurement of the release of cytoplasmic constituents). The peptides were added at concentrations of 10 or 100 μmol L^{-1}. The concentration of free K^+ in the suspensions and absorbance values at 260 nm in the supernatant were measured after treatment at 0, 3, 6, 9, 12, 24, and 48 h. Each treatment was performed in triplicate, and the control lacked peptides.

2.6. The Fungicidal Kinetics of PAF56 against Spores

The fungicidal kinetics of PAF56 against *P. digitatum*, *P. italicum*, and *G. candidum* spores were measured as previously described [5,16] with minor modifications. The final concentration of PAF56 was used at 64 μmol L^{-1}. Spores (10^3 CFU mL^{-1}) were mixed with PAF56 in sterile distilled water. Group without the use of peptide was the control group. The spore suspension was then incubated at 25 °C. Samples of 50 μL were spread onto PDA plates at each time point after incubation. The CFU was counted after the plates were incubated for 48 h at 25 °C. Treatments were prepared in triplicate.

2.7. Damage Effect of PAF56 on Membrane Permeability of Spores by Fluorescence Microscopy

Damage effect of PAF56 on membrane permeability of the three fungi spores was observed by an Eclipse TS100 epifluorescence microscope (Nikon Corporation, Minato City, Japan), and the fluorescent dye SYTOX Green (SG) (Molecular Probes; Invitrogen Corp, Carlsbad, CA, USA) was used in this experiment according to the previous description with minor modifications [10,17]. Aliquots of 450 μL of spores (10^7 CFU mL^{-1}) were incubated in 1.5 mL light-safe microcentrifuge tubes, and subsequently, 50 μL of PAF56 (the final concentration was 64 μmol L^{-1}) was added and allowed to grow. Group without the use of peptide was the control group. The suspensions of fungal spores were stained with fluorescent dye SYTOX Green after incubation. Fluorescence was examined and photographed with FITC filter sets at different time points. We also captured the simultaneous brightfield images.

Confocal microscopy was used to observe the effect of PAF56 on spores as well. Since the spores of *G. candidum* are bigger than those of *P. digitatum* and *P. italicum*, it is easier to observe. Thus, this experiment took *G. candidum* as an example. Confocal microscopy was used to observe the distribution of fluorescent-labeled TMR-PAF56 in spores and the damage to cell membrane at different times. Aliquots of 450 μL of spores (10^7 CFU mL^{-1}) were incubated in 1.5 mL light-safe microcentrifuge tubes. Subsequently, 50 μL of TMR-PAF56 (the final concentration was 64 μmol L^{-1}) and fluorescent dye SG (the final concentration was 0.2 μmol L^{-1}) were added. The suspensions were photographed by an Olympus FV1000 laser confocal microscope (Olympus Corporation, Shinjuku City, Japan) with FITC filter and Rhodamine Red-x sets. Simultaneous brightfield images were also photographed.

3. Results

3.1. Analysis of PAF56's Effect on the Induction of Defense Resistance in Citrus Fruit

The fruit test experiment was carried out to judge whether PAF56 could induce the defense resistance of the citrus fruit by inoculating PAF56 into the citrus fruit in the different wounds (Figure 1). The disease incidence and lesion diameter results show that there were no significant difference between the citrus fruit inoculated with PAF56 and the control group inoculated with sterile distilled water during the whole period of the disease development. The results for the three kinds of pathogenic fungi, *P. italicum*, *P. digitatum*, and *G. candidum*, were similar, which demonstrates that PAF56 might not induce defense resistance of citrus fruit.

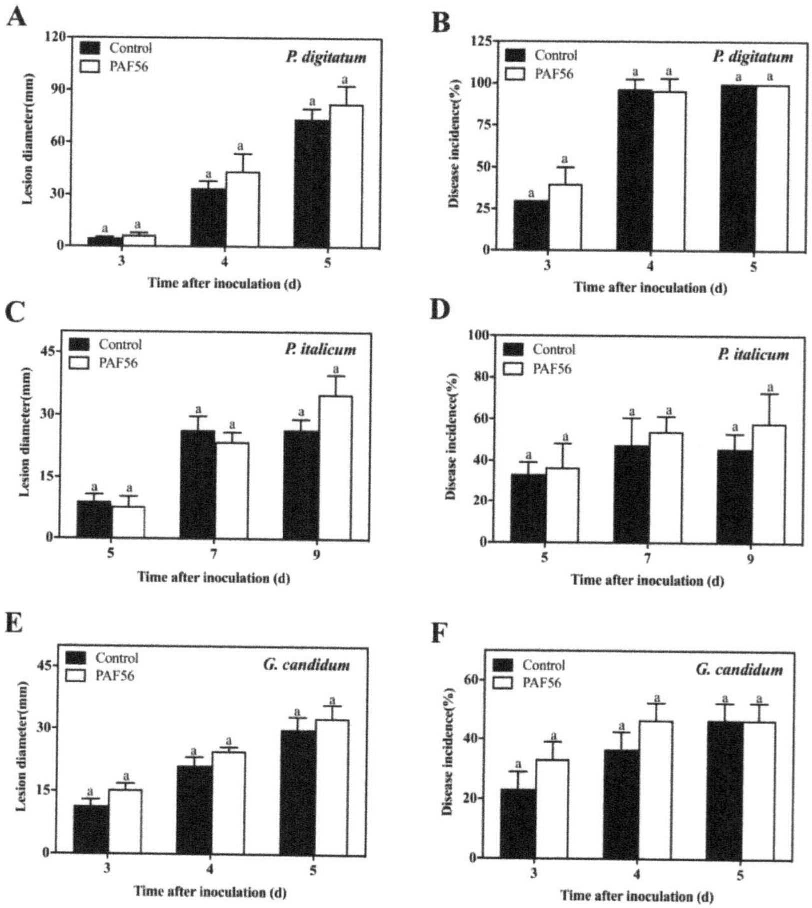

Figure 1. Effect of PAF56 (Inoculation in the different wounds) on lesion diameter and disease incidence of citrus fruit caused by *P. digitatum* (**A**,**B**), *P. italicum* (**C**,**D**), and *G. candidum* (**E**,**F**). Values are mean ± SD. The letters 'a' indicate no differences at the 0.05 level. The analysis was conducted using the data from the same pathogen on the same day.

3.2. Morphological Alterations of Fungal Mycelia in Response to PAF56

The effect of PAF56 on the morphology of the three fungi mycelia was examined with SEM (Figure 2). It could be observed that the mycelial morphology of the two concentrations of the treated groups changed considerably compared with the control group. The mycelia treated with PAF56 of 10 or 100 μmol L^{-1} became deformed, shrunken, and distorted, and higher concentrations showed higher damage. By comparison, the surface morphology of *P. digitatum* hyphae changed most obviously after being treated by PAF56 at 100 μmol L^{-1}. Meanwhile, the mycelia were seriously collapsed, distorted, and fibrous.

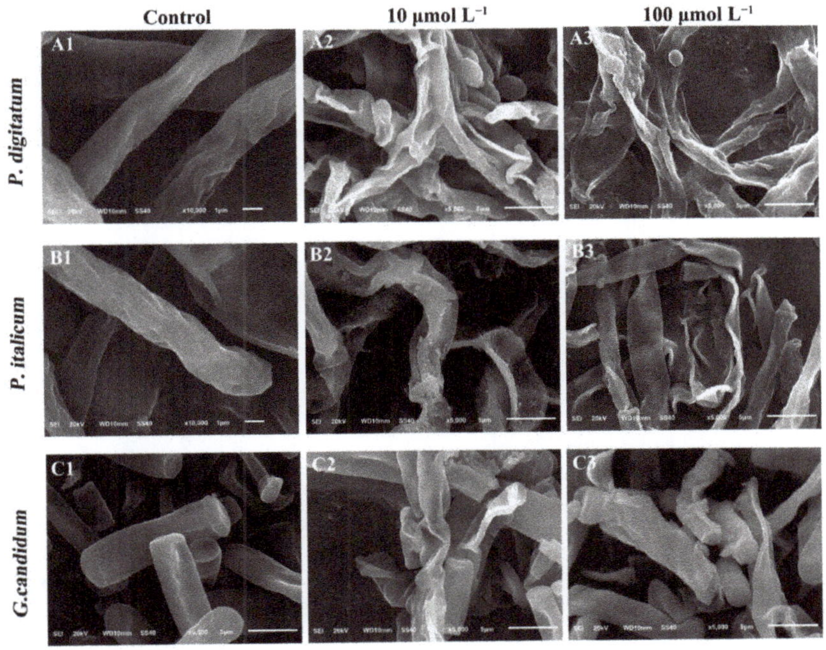

Figure 2. SEM images of *P. digitatum* (panels **A**), *P. italicum* (panels **B**), and *G. candidum* (panels **C**) mycelia treated with PAF56. Mycelia were incubated in 5% PDB without PAF56 (**A1–C1**) or with PAF56 at final concentrations of 10 μmol L^{-1} (**A2–C2**) or 100 μmol L^{-1} (**A3–C3**).

3.3. Effect of PAF56 on the Efflux of K^+ and the Release of Cytoplasmic Constituents of Mycelia

The potassium ions (K^+) leakage of mycelia was caused by PAF56 treatment (Figure 3A,C,E). As shown, the concentration of extracellular K^+ in the group treated with PAF56 and the control group increased gradually during the measurements, and PAF56 treatment could significantly induce the efflux of K^+. The K^+ leakage of the PAF56 group with the concentration of 100 μmol L^{-1} was significantly higher ($p < 0.05$) than that of the control (no peptide) group. Specifically, *G. candidum* mycelia was seriously damaged at high concentration (100 μmol L^{-1}), which led to the leakage of K^+. After the treatment with peptides, the concentration of the extracellular K^+ could be reached at 3 h (Figure 3E). This may be related to the characteristics of *G. candidum*, the cell membrane of which is easier for PAF56 to break.

The release of cytoplasmic constituents from the treated mycelia of the three fungi was measured as well. To analyze the leakage of cytoplasmic constituents in the cells of these pathogenic fungi mycelia after being treated with PAF56, the OD_{260} value was measured. The results show that the OD_{260} value of the PAF56 treatment group revealed an increasing release of cytoplasmic constituents with the exposure time increasing (Figure 3B,D,F), compared with that of the control group. This means that the release of cytoplasmic constituents results from these pathogenic fungi mycelia being treated with PAF56.

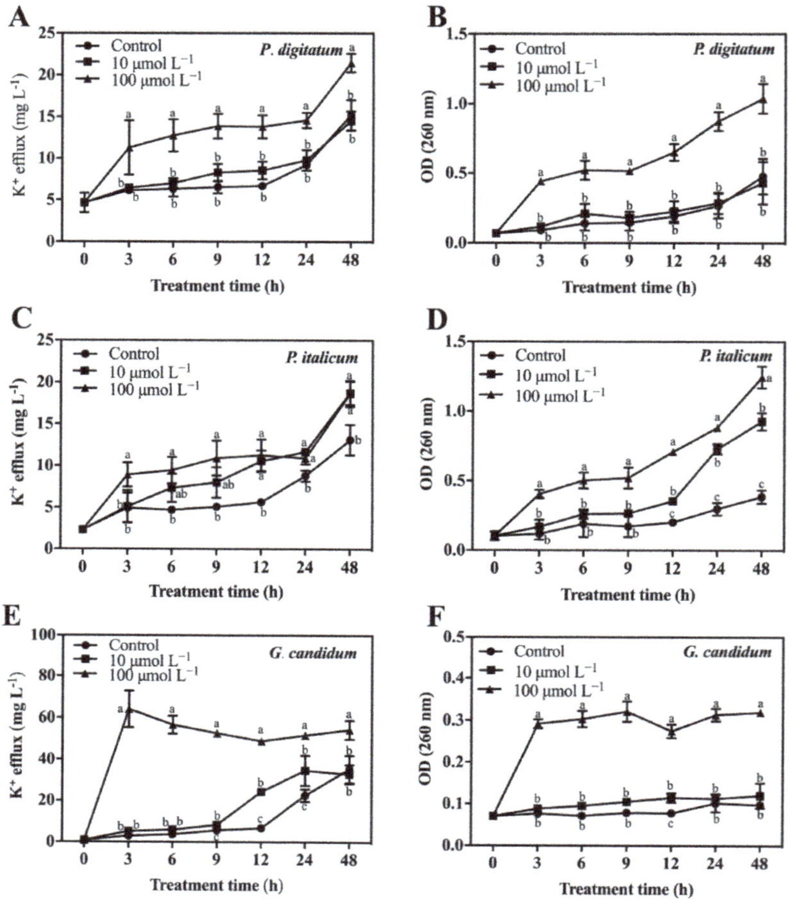

Figure 3. K^+ efflux and release of cellular constituents of *P. digitatum* (**A**,**B**), *P. italicum* (**C**,**D**), and *G. candidum* (**E**,**F**) mycelia treated with PAF56. Mycelia were incubated in 10 μmol L^{-1} or 100 μmol L^{-1} or without PAF56 (control) solutions. The mycelia were washed before resuspension in sterilized distilled water (for the measurement of extracellular potassium concentration) or phosphate buffer (0.05 mol L^{-1}, phosphate, pH 7.0) (for the measurement of the release of cytoplasmic constituents). The concentration of free K^+ in the suspensions without mycelia was measured by flame atomic absorption spectroscopy, and the release of cytoplasmic constituents was measured using a Multiskan Spectrum microplate spectrophotometer at 260 nm. Vertical bars indicate the standard error of the means. The letters 'a', 'b', and 'c' indicate significant differences at the 0.05 level.

3.4. *Effect of PAF56 Treatment on the Membrane Permeability of Fungal Spores*

The effects of PAF56 treatment on the membranes of the three fungi spores were observed by SG fluorescent staining and fluorescence microscopy. Figure 4A,C,E are the images of the bright field of Figure 4B,D,F. The results show that no spores in the control group could emit green SG fluorescence (Figure 4B1–B3). PAF56 could not induce the spores to emit green SG fluorescence in a short time (3–5 min) (Figure 4D1–D3) either. However, PAF56 treatment after 16 h could cause the emitting of green SG fluorescence for all spores (Figure 4F1–F3). To sum up, PAF56 treatment could change the spore membrane permeability of *P. digitatum*, *P. italicum*, and *G. candidum*, allowing SG to enter the spores and bind with the nucleic acid. These results are probably related to the duration of the treatment with PAF56.

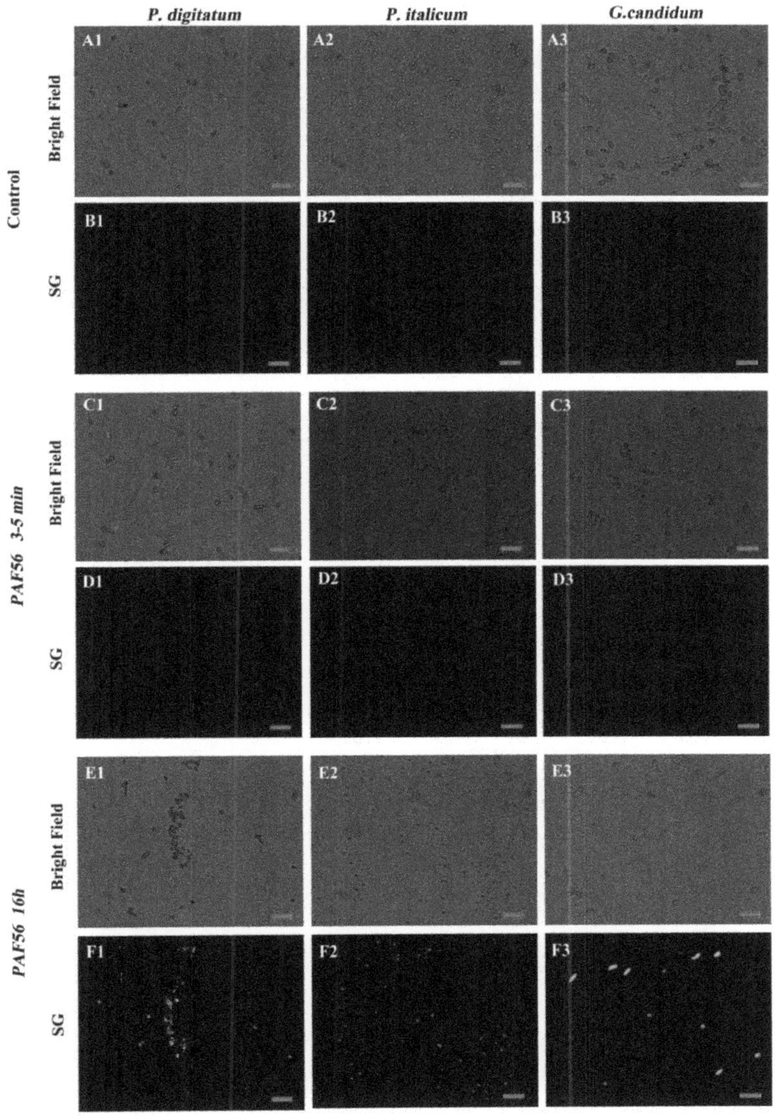

Figure 4. Effect of PAF56 treatment on the membrane permeability of *P. digitatum* (**A1–F1**), *P. italicum* (**A2–F2**), and *G. candidum* (**A3–F3**) spores (bars = 20 μm). Spores were incubated in light-safe microcentrifuge tubes, and subsequently, PAF56 (the final concentration was 64 μmol L^{-1}) was added. Group without the use of peptide was the control group. The suspensions of fungal spores were stained with fluorescent dye SYTOX Green (SG) after incubation, and then fluorescence was examined and photographed at different time points.

3.5. Time-Kill Kinetics of PAF56 against Fungal Spores

PAF56 was incubated with the three pathogenic fungi, and the time–kill kinetics curves of PAF56 were plotted at different times by measuring the number of colonies (Figure 5). Compared with the control group, the CFU number of *P. digitatum*, *P. italicumi*, and *G. candidum* decreased with incubation time after adding PAF56 to all treatment groups. The effect of PAF56 on *P. digitatum* and *P. italicum* spores was significantly related to the action time. The longer the action time was, the lower the detected CFU number. The effect of PAF56 on *G. candidum* spores was very strong in a short treatment time. The number of colonies of *G. candidum* was 0 at 3 min. This might be related to the differences in the structure of spores of the three pathogens and their sensitivity to PAF56.

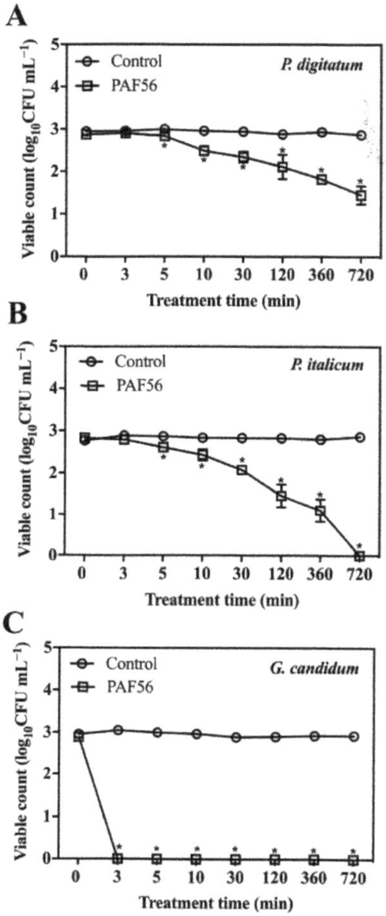

Figure 5. Time–kill kinetics of peptides PAF56 against *P. digitatum* (**A**), *P. italicum* (**B**), and *G. candidum* (**C**) spores. Spores (10^3 CFU mL^{-1}) were mixed with PAF56 (64 μmol L^{-1}) in sterile distilled water. Group without the use of peptide was the control group. Samples of 50 μL were spread onto PDA plates at each time point after incubation. The CFU was counted after the plates were incubated for 48 h at 25 °C. Vertical bars indicate the standard error of the means. The mark * represents the significant differences ($p < 0.05$) between PAF56 and the control group.

3.6. Time-Lapse Confocal Fluorescence Microscopy Analyses of the Interaction of TMR-PAF56

As presented in Figure 6, TMR-PAF56 could emit red fluorescence after excitation, and the red fluorescence first appeared on the spore surface of *G. candidum*. Then, the red fluorescence was detected in the spores, which indicated that TMR-PAF56 firstly gathered on the surfaces of the spores, then slowly entered the spores as the time extended, and finally spread to the insides of the whole spores. At the same time, the green fluorescence of SG was detected. The green SG fluorescence was not observed at the beginning of 20 min but appeared at the following time points. Almost all spores produced a strong green

SG fluorescence at 100 min, which indicated TMR-PAF56 destroyed the spore membrane. The results were that SG entered the spores, bound with cytoplasmic constituents, and emitted green SG fluorescence. The appearance and increase of green SG fluorescence corresponded to the degree of damage of TMR-PAF56 to the cell membrane. In summary, the damage degree of TMR-PAF56 to the membrane of *G. candidum* spore is closely related to the acting time, and the main target of TMR-PAF56 may be the membrane of spores.

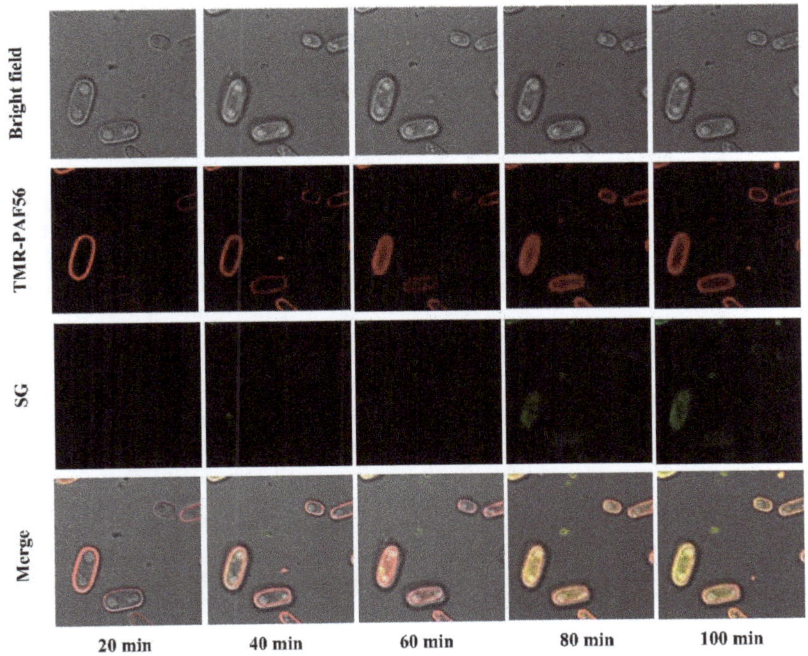

Figure 6. Time-lapse confocal fluorescence microscopy analyses of the interaction of TMR-PAF56. Spores were incubated in light-safe microcentrifuge tubes. Subsequently, TMR-PAF56 and fluorescent dye SG were added. The suspensions were photographed by an Olympus FV1000 laser confocal microscope with FITC filter and Rhodamine Red-x sets at different time points. Simultaneous brightfield images were also photographed.

4. Discussion

It is important to study the mechanism of peptides' effects on pathogenic fungi for their better application in the future. As a safe peptide without hemolysis [10], it is necessary to conduct in-depth research on PAF56.

Current studies on the controlling of fruit disease have found that a few substances can not only directly inhibit fungi, but also induce fruit resistance to disease. The incidence of disease was significantly reduced, such as antagonistic yeast [18], salicylic acid [19], and chitosan oligosaccharide [20–22]. However, few studies have reported whether the peptides can induce fruit resistance, while only some researchers thought that the peptides could stimulate the immune response and function in vivo [23,24]. Therefore, in this study, PAF56 and pathogenic spores were inoculated in different holes of citrus fruit to verify whether PAF56 could induce fruit resistance. The results show that PAF56 did not induce disease resistance in citrus fruit (Figure 1). This indicated that the control of green and blue mold and acid rot of citrus fruit was due to the direct action of PAF56 on pathogenic fungi. Therefore, we further investigated the mechanism of the peptide PAF56 on pathogenic fungi in vitro.

The cell membrane represents the first and last line of defense for ensuring the normal function and ultimately the viability of the cell. The surface morphological changes of mycelia treated with PAF56 were observed by scanning electron microscope. The results show that PAF56 caused mycelia wrinkles, irregular distortion, and serious morphology changes in *P. digitatum*, *P. italicum*, and *G. candidum* after 16 h inoculation (Figure 2). Those results were similar to those found with Buforin 2, some essential oils, and citral [12,25,26].

Previous studies showed that the value of the extracellular conductivity of the three fungi would increase. We further measured the leakage of K^+ and the release of cytoplasmic constituents of fungal mycelia after treatment with PAF56 (Figure 3). These three indicators are often used to assess irreversible damage to cell membranes and cytoplasm. The experimental results further confirm that PAF56 could increase the permeability of the cell membranes of the three fungi.

P. digitatum, *P. italicum*, and *G. candidum* initiate their infection of citrus fruit by disseminating spores. It is extremely important to control spore germination and infection. The destruction of spores' membranes in *P. digitatum*, *P. italicum*, and *G. candidum* treated with PAF56 over short term (3–5 min) and long term (16 h) was observed by SG fluorescence staining technology and fluorescence microscope (Figure 4). The results show that the spore membrane of these three pathogenic fungi could be destroyed after treatment with PAF56 for 16 h. It is noteworthy that the treatment time of 3–5 min is insufficient to destroy the spore membrane of the three pathogens. Furthermore, the kinetic curves of spores killed by PAF56 were plotted (Figure 5). It is also noteworthy that the effect of PAF56 on *P. digitatum* and *P. italicum* spores became more and more remarkable with the prolongation of time, and the kinetic curves were almost linear, indicating that the effect of PAF56 on *P. digitatum* and *P. italicum* spores was positively correlated with treatment time. The difference is that the effect of PAF56 on *G. candidum* spores was very significant in a short period of time, which might be related to the difference in the structure of spores of pathogenic fungus and the differing sensitivity of spores to PAF56. Then, PAF56 was labeled by fluorescence labeling, and the localization of TMR-PAF56 in *G. candidum* spores was observed by laser confocal microscopy at different times (Figure 6). The results show that the damage degree of TMR-PAF56 on *G. candidum* spores was closely related to the action time, which is similar to previous studies [27,28]. Fluorescent labeling of PAF56 might lead to a decrease in its antifungal activity. The effect of PAF56 on spores was directly related to the destruction of the membrane.

5. Conclusions

In summary, PAF56 treatment could destroy the cell structure of *P. digitatum*, *P. italicum*, and *G. candidum* mycelia, change their permeability, and cause leakage of their contents. Citrus fruit could not be induced by PAF56 to produce disease resistance. The effect of PAF56 on spores is directly related to the breaking of cell membranes and the acting time. The results of this study will provide a useful reference for related studies and applications in citrus production and storage.

Author Contributions: Conceptualization, K.Z.; methodology, W.W. and G.F.; software, W.W.; validation, W.W.; formal analysis, G.F. and X.L.; investigation, W.W.; resources, K.Z.; data curation, J.M.; writing—original draft preparation, W.W. and C.R.; writing—review and editing, W.W. and G.F.; visualization, X.L.; supervision, J.M. and C.R.; project administration, K.Z.; funding acquisition, K.Z. All authors have read and agreed to the published version of the manuscript.

Funding: This research was supported by the National Natural Science Foundation of China (Grant No. 31972126).

Data Availability Statement: Not applicable.

Conflicts of Interest: The authors declare no conflict of interest.

References

1. Ciociola, T.; Giovati, L.; Conti, S.; Magliani, W.; Santinoli, C.; Polonelli, L. Natural and synthetic peptides with antifungal activity. *Futur. Med. Chem.* **2016**, *8*, 1413–1433. [CrossRef]
2. Keymanesh, K.; Soltani, S.; Sardari, S. Application of antimicrobial peptides in agriculture and food industry. *World J. Microbiol. Biotechnol.* **2009**, *25*, 933–944. [CrossRef]
3. Jenssen, H.; Hamill, P.; Hancock, R. Peptide Antimicrobial Agents. *Clin. Microbiol. Rev.* **2006**, *19*, 491–511. [CrossRef]
4. López-García, B.; Pérez-Payá, E.; Marcos, J.F. Identification of Novel Hexapeptides Bioactive against Phytopathogenic Fungi through Screening of a Synthetic Peptide Combinatorial Library. *Appl. Environ. Microbiol.* **2002**, *68*, 2453–2460. [CrossRef]
5. Muñoz, A.; Lopez-Garcia, B.; Marcos, J.F. Comparative Study of Antimicrobial Peptides to Control Citrus Postharvest Decay Caused byPenicillium digitatum. *J. Agric. Food Chem.* **2007**, *55*, 8170–8176. [CrossRef]
6. López-García, B.; Veyrat, A.; Pérez-Payá, E.; González-Candelas, L.; Marcos, J.F. Comparison of the activity of antifungal hexapeptides and the fungicides thiabendazole and imazalil against postharvest fungal pathogens. *Int. J. Food Microbiol.* **2003**, *89*, 163–170. [CrossRef]
7. Muñoz, A.; López-García, B.; Marcos, J.F. Studies on the Mode of Action of the Antifungal Hexapeptide PAF26. *Antimicrob. Agents Chemother.* **2006**, *50*, 3847–3855. [CrossRef] [PubMed]
8. López-García, B.; Ubhayasekera, W.; Gallo, R.L.; Marcos, J.F. Parallel evaluation of antimicrobial peptides derived from the synthetic PAF26 and the human LL37. *Biochem. Biophys. Res. Commun.* **2007**, *356*, 107–113. [CrossRef]
9. López-García, B.; Harries, E.; Carmona, L.; Campos-Soriano, L.; López, J.J.; Manzanares, P.; Gandía, M.; Coca, M.; Marcos, J.F. Concatemerization increases the inhibitory activity of short, cell-penetrating, cationic and tryptophan-rich antifungal peptides. *Appl. Microbiol. Biotechnol.* **2015**, *99*, 8011–8021. [CrossRef] [PubMed]
10. Wang, W.; Deng, L.; Yao, S.; Zeng, K. Control of green and blue mold and sour rot in citrus fruits by the cationic antimicrobial peptide PAF56. *Postharvest Biol. Technol.* **2018**, *136*, 132–138. [CrossRef]
11. Jeong, R.-D.; Chu, E.-H.; Lee, G.W.; Cho, C.; Park, H.-J. Inhibitory effect of gamma irradiation and its application for control of postharvest green mold decay of Satsuma mandarins. *Int. J. Food Microbiol.* **2016**, *234*, 1–8. [CrossRef]
12. Tao, N.; OuYang, Q.; Jia, L. Citral inhibits mycelial growth of Penicillium italicum by a membrane damage mechanism. *Food Control* **2014**, *41*, 116–121. [CrossRef]
13. Droby, S.; Vinokur, V.; Weiss, B.; Cohen, L.; Daus, A.; Goldschmidt, E.E.; Porat, R. Induction of Resistance to Penicillium digitatum in Grapefruit by the Yeast Biocontrol Agent Candida oleophila. *Phytopathology* **2002**, *92*, 393–399. [CrossRef] [PubMed]
14. Bajpai, V.K.; Sharma, A.; Baek, K.-H. Antibacterial mode of action of Cudrania tricuspidata fruit essential oil, affecting membrane permeability and surface characteristics of food-borne pathogens. *Food Control* **2013**, *32*, 582–590. [CrossRef]
15. Paul, S.; Dubey, R.; Maheshwari, D.K.; Kang, S.C. Trachyspermum ammi (L.) fruit essential oil influencing on membrane permeability and surface characteristics in inhibiting food-borne pathogens. *Food Control* **2010**, *22*, 725–731. [CrossRef]
16. Li, L.; Shi, Y.; Cheserek, M.J.; Su, G.; Le, G. Antibacterial activity and dual mechanisms of peptide analog derived from cell-penetrating peptide against Salmonella typhimurium and Streptococcus pyogenes. *Appl. Microbiol. Biotechnol.* **2012**, *97*, 1711–1723. [CrossRef]
17. Puig, M.; Moragrega, C.; Ruz, L.; Calderón, C.E.; Cazorla, F.M.; Montesinos, E.; Llorente, I. Interaction of antifungal peptide BP15 with Stemphylium vesicarium, the causal agent of brown spot of pear. *Fungal Biol.* **2016**, *120*, 61–71. [CrossRef]
18. Spadaro, D.; Droby, S. Development of biocontrol products for postharvest diseases of fruit: The importance of elucidating the mechanisms of action of yeast antagonists. *Trends Food Sci. Technol.* **2016**, *47*, 39–49. [CrossRef]
19. Janda, T.; Pál, M.; Darkó, É.; Szalai, G. Use of Salicylic Acid and Related Compounds to Improve the Abiotic Stress Tolerance of Plants. *Pract. Asp.* **2017**, 35–46. [CrossRef]
20. Meng, X.; Yang, L.; Kennedy, J.F.; Tian, S. Effects of chitosan and oligochitosan on growth of two fungal pathogens and physiological properties in pear fruit. *Carbohydr. Polym.* **2010**, *81*, 70–75. [CrossRef]
21. Zhao, X.; She, X.; Du, Y.; Liang, X. Induction of antiviral resistance and stimulary effect by oligochitosan in tobacco. *Pestic. Biochem. Physiol.* **2007**, *87*, 78–84. [CrossRef]
22. Yin, H.; Zhao, X.; Du, Y. Oligochitosan: A plant diseases vaccine—A review. *Carbohydr. Polym.* **2010**, *82*, 1–8. [CrossRef]
23. Tossi, A.; Sandri, L.; Giangaspero, A. Amphipathic, α-helical antimicrobial peptides. *Pept. Sci.* **2015**, *55*, 4–30. [CrossRef]
24. Reddy, K.V.R.; Yedery, R.D.; Aranha, C. Antimicrobial peptides: Premises and promises. *Int. J. Antimicrob. Agents* **2004**, *24*, 536–547. [CrossRef] [PubMed]
25. Hao, G.; Shi, Y.-H.; Tang, Y.-L.; Le, G.-W. The membrane action mechanism of analogs of the antimicrobial peptide Buforin 2. *Peptides* **2009**, *30*, 1421–1427. [CrossRef] [PubMed]
26. Helal, G.A.; Sarhan, M.M.; Abu Shahla, A.N.K.; Abou El-Khair, E.K. Effects of Cymbopogon citratus L. essential oil on the growth, morphogenesis and aflatoxin production of Aspergillus flavus ML2-strain. *J. Basic Microbiol.* **2007**, *47*, 5–15. [CrossRef]
27. Madani, F.; Lindberg, S.; Langel, U.; Futaki, S.; Gräslund, A. Mechanisms of Cellular Uptake of Cell-Penetrating Peptides. *J. Biophys.* **2011**, *2011*, 414729. [CrossRef]
28. Henriques, S.T.; Melo, M.N.; Castanho, M.A.R.B. Cell-penetrating peptides and antimicrobial peptides: How different are they? *Biochem. J.* **2006**, *399*, 1–7. [CrossRef]

MDPI
St. Alban-Anlage 66
4052 Basel
Switzerland
Tel. +41 61 683 77 34
Fax +41 61 302 89 18
www.mdpi.com

Foods Editorial Office
E-mail: foods@mdpi.com
www.mdpi.com/journal/foods

www.ingramcontent.com/pod-product-compliance
Lightning Source LLC
Chambersburg PA
CBHW061629210926
51850CB00009B/2489
9783036564753